VIRTUE ETHICS IN SOCIAL WORK PRACTICE

Using evidence from research with practitioners, integrated with wider material about virtue ethics in the helping professions, this book explores important types of virtue that are central to developing and sustaining excellence in social work.

Comprised of ten chapters and drawing on extensive research with social workers as well as wider debates and analysis, the discussion carefully concentrates on everyday experiences and achievements. This approach enables the book to avoid an idealized and prescriptive approach by making clear that virtues vary between contexts and individuals, while at the same time clearly marking out qualities and characteristics of social work that are foundational to the development of practitioners and of the profession as a whole.

It will be required reading for students on all BSc/BSW and MSc/MSW courses on professional ethics or preparation for practice. It will also be of interest to practitioners in other professions, including human services, health, education and social development or development studies.

Richard Hugman is Professor of Social Work at the University of New South Wales, Sydney, Australia. He originally qualified and practised as a social worker in the UK before migrating to Australia, where he has taught and researched at several universities and continued some practice. From 2002 to 2014 he was a member of the International Federation of Social Workers Committee on Ethics, chairing it for six years; he has also been a member of the Asia-Pacific regional committee of the International Federation of Social Workers. Richard has undertaken research and consultancy with international organizations, including with UNICEF Vietnam. He has published extensively about social work and social work ethics, including the book *Culture, Values and Ethics in Social Work* (Routledge, 2013).

Manohar Pawar is Professor of Social Work at the School of Humanities and Social Sciences, Charles Sturt University, Australia, and is President of the International Consortium for Social Development. He has over 35 years of experience in social work education, research and practice in Australia and India. He is the lead chief investigator of research funded by the Australian Research Council's Discovery project that focuses on virtues and social work practice, which inspired this book. Manohar's interests and publications include *Community Development in Asia and the Pacific* (Routledge, 2010) and *Social Development* (co-edited, Routledge, 2010).

A. W. (Bill) Anscombe is Adjunct Associate Professor with the Australian Centre for Christianity and Culture at Charles Sturt University, Australia, with a long career in social work practice (predominantly corrections and child protection), social work education and research. His practice has included senior positions as Operations Manager for New South Wales (Southern) Corrections and Director of Child and Family Services (Western) in the Australian state of New South Wales. He is currently a voluntary director of five not-for-profit human service companies or associations. Bill's research interests have been broad, including Indigenous housing and governance, multicultural Australia, rural social services and the rural Church. He has undertaken many evaluation projects ranging from chaplaincies, early-childhood services, disability through to multicultural health.

Amelia Wheeler is a Lecturer in Social Work at the Australian Catholic University. Amelia previously practised in the fields of child protection, youth homelessness and family law and has taught and researched at the University of New South Wales. Her recently completed PhD involved an ethical analysis of child-inclusive practices in Australian post-separation family mediation. Amelia has held research positions with the Gendered Violence Research Network at UNSW and the Australian Institute of Relationship Studies, and has been engaged in research consultancies with a number of non-government organisations. Her research interests include domestic and family violence, child and youth participation in social services, anti-racist social work and the use of ethics in everyday professional life.

VIRTUE ETHICS IN SOCIAL WORK PRACTICE

Richard Hugman, Manohar Pawar,
A. W. (Bill) Anscombe and Amelia Wheeler

Routledge
Taylor & Francis Group
LONDON AND NEW YORK

First published 2021
by Routledge
2 Park Square, Milton Park, Abingdon, Oxon OX14 4RN

and by Routledge
52 Vanderbilt Avenue, New York, NY 10017

Routledge is an imprint of the Taylor & Francis Group, an informa business

British Library Cataloguing-in-Publication Data
A catalogue record for this book is available from the British Library

Library of Congress Cataloging-in-Publication Data
Names: Hugman, Richard, 1954– author. | Anscombe, A. W., author. | Pawar, Manohar S., author. | Wheeler, Amelia, author.
Title: Virtue ethics in social work practice / Richard Hugman, Manohar Pawar, A.W. (Bill) Anscombe and Amelia Wheeler.
Description: Abingdon, Oxon ; New York, NY : Routledge, 2020. | Includes bibliographical references and index.
Identifiers: LCCN 2020012344 (print) | LCCN 2020012345 (ebook) | ISBN 9780367429447 (hbk) | ISBN 9780367434342 (pbk) | ISBN 9781003003151 (ebk)
Subjects: LCSH: Social service–Moral and ethical aspects. | Social workers–Professional ethics.
Classification: LCC HV10.5 .H845 2020 (print) | LCC HV10.5 (ebook) | DDC 174/.936132–dc23
LC record available at https://lccn.loc.gov/2020012344
LC ebook record available at https://lccn.loc.gov/2020012345

ISBN: 978-0-367-42944-7 (hbk)
ISBN: 978-0-367-43434-2 (pbk)
ISBN: 978-1-003-00315-1 (ebk)

Typeset in Bembo
By Wearset Ltd, Boldon, Tyne and Wear

CONTENTS

PREFACE

In social work ethics, issues of respect for people, human rights and social justice are widely debated and discussed. However, there is a third pillar to professional ethics in social work that is recognised in the international statement of ethical principles shared by the International Federation of Social Workers and the International Association of Schools of Social Work, namely virtue ethics. Often referred to in national codes of ethics in terms of professional integrity, virtue ethics addresses the ethical qualities of social workers as these are embedded in their practices. Yet there is considerably less attention to virtue ethics in social work writing and research than there is to the other pillars of respect and rights and social justice. This book originates in a project to investigate virtue ethics in social work by bringing together analysis of what social workers themselves say about their practice and the ethical qualities that they see as important, with a deeper discussion of the ethical ideas that underpin this analysis.

This current volume is one key part of a wider project to explore virtue ethics in social work, focusing particularly on practice. It is focused on one key phase of a wider project. The reader who is interested in considering the other parts of this overall work may look at Pawar, Hugman, Alexandra and Anscombe (eds) (2017) *Empowering Social Workers: Virtuous Practitioners* (Springer) for the presentation of ten biographies of 'ethical exemplars' in Australian social work, and Pawar, Hugman, Anscombe and Alexandra (2019) 'Searching for virtue ethics: a survey of social work ethics curriculum and educators', *British Journal of Social Work*, DOI: 10.1093/bjsw/bcz106 for a discussion of virtue ethics in social work education programmes. At the same time, each of these discussions stands alone in addressing specific issues and sets of ideas in the field.

This discussion is presented with the hope that virtue ethics will be seen by social workers more clearly as a central concern. In addition to being concerned with the ethical goals of social work, correctly in our view, practitioners also need to be attentive to the ways in which they practice and the ethical qualities that they bring to this practice as part of the ethical ground of the profession. Our experience has been that through this process we have found our colleagues to be interested in exploring these questions, so our goal is that the discussion presented here will contribute to wider recognition of the importance of virtue ethics as well as familiarity with the ideas and terms that enable this conversation to develop.

ACKNOWLEDGEMENTS

A book such as this would not be possible without contribution from many quarters. Crucial to the discussion presented here are the ideas and accounts provided by social work colleagues from all the states and territories in Australia. We would like to acknowledge the time they gave to this project, and the insights and experiences they shared with us.

We are also very grateful to other colleagues in Australia and around the world, whose contributions include discussions of general ideas, specific critiques and practical assistance. In particular we would like to acknowledge Sarah Banks, Wendy Bowles, Vivienne Bozalek, Simone Engdahl, Kris Gibbs, Mel Gray, Dorothee Holscher, Sacha Kendall, Debra Noy, Jessica Russ-Smith, Nikki Scott. In particular, Heather Kelly provided skilful editing of the final manuscript, which improved the clarity of presentation of these ideas and debates; Angela Lowe-Liddell created the cover design.

This research was supported under the Australian Research Council's Discovery Project funding scheme (DP140103730; project title, *Virtuous Practitioners: Empowering Social Workers*) and we are grateful to all those who have contributed to this decision. We are also thankful to Charles Sturt University's Human Research Ethics Committee for approving the research proposal (protocol no: 2014/057) and for support from the Research Office of the University of New South Wales.

Finally, at Routledge, Claire Jarvis, Catherine Jones and their team have provided support and assistance at all stages in producing this book. We sincerely thank them.

Although this book is a product of excellent teamwork, we authors are responsible for any shortcomings and limitations in it.

1

INTRODUCTION TO VIRTUE ETHICS

> **CHAPTER OBJECTIVES**
>
> - To outline the purpose and structure of the book
> - To explore virtue ethics as a ground for professional ethics, and to outline its place in social work practice

Professional ethics as a foundation for social work

This book explores the relevance of virtue ethics to social work and the ways in which virtues can assist in understanding and developing good practice. Our interest in this aspect of professional ethics comes from recognising that virtue ethics is addressed much less frequently than the key principles of human rights and social justice. Our goal is to extend the range of thinking about professional ethics in social work and so to add to the ways in which ethics can inform the development of practice.

Social work has long placed great importance on values as key to defining its goals and methods (Timms 1983). Furthermore, some analyses go further and describe social work as a 'values-based' profession (Bisman 2004; Chechak 2015: 41–43; Dimitrijoska and Ilievski 2016: 51–52; Reamer 2018: 3). This can be seen in the international definition of social work, which states that '[p]rinciples of social justice, human rights, collective responsibility and respect for diversities are central to social work' (IFSW/IASSW 2018: n.p.). For this reason, it is vital that social workers are able to identify and consider critically the values of their work, in order to grasp and to shape its ends and means and the relationship between them. It is this process of identifying and considering values that we understand as ethics

(Hinman 2013: 5). This understanding gives ethics a central place as the foundation of social work's purpose and the forms of practice that make up the profession as a whole.

Banks explains values as 'particular types of beliefs that people hold about what is regarded as worthy or valuable' (2012: 8). Banks points out that these include personal and social values, and may be drawn from religious or political beliefs or from philosophy. For philosophers such as MacIntyre (1989) or Hinman (2013), the values that ethics considers are about the purposes and meanings people ascribe to their lives; that is, ethics is concerned with questions of right and wrong, or good and bad in human conduct and character, or in social relationships and systems. This is quite distinct from other forms of value, such as economic valuation of goods, which may be referred to as 'non-moral values'. In this way we can see that ethics is concerned with moral values, noting that such an understanding is much broader than the interpretation of morality that sees it as a concern with dogmatic prescription regarding specific human behaviours (such as sexual relationships). Rather, such a definition of morality draws on the ancient Roman term 'mores', referring to the deep understanding of values across cultures and societies that affect all aspects of life.

In this sense, ethics is a considered framework that provides guidelines to what is good and what is right in human life. From this, we can say that ethics in social work is focused on the conduct and character of social workers in the context of the social relationships and systems within which it is practised. Ethics has become an increasing concern within social work in recent times, and now forms an identifiably specialist strand within the literature that has grown rapidly. Some social work ethicists, such as Reamer (2001), regard this as part of the maturation of social work as a profession. Certainly in this period social work has gained a stronger identity in many countries, in being recognised more by governments, other professions and the wider society. This includes engagement with evidence-based and evidence-informed practice and the refinement of methods and techniques (for example, Smith 2004; Thyer 2013; Plath 2014), as well as more robust debate about issues such as the exercise of power (for example, Hugman 1991; Rees 1991; Clifford and Burke 2009; Pullen-Sansfaçon and Cowden 2013) and colonial and post-colonial dynamics on the development of social work internationally (for example, Gray 2005; Bennett and Green 2019).

From a somewhat different perspective, Banks (2012: xix), herself a major contributor to the process of expanding analysis of social work ethics, notes the way in which this has happened as part of a wider growth of interest in ethics in public life and even more generally in the wider society. Banks links the increasingly broad interest in ethics to perceptions of wrongdoing in public life and of intractable social problems (currently often referred to as 'wicked' problems), for which standards need to be established and asserted. For this reason, the increased interest in ethics shown by social work is part of a wider social concern to ensure that occupations that exercise profound influence, even control, over people's lives act with responsibility. This is achieved, at least partly, through frameworks of accountability that are created by formal, publicly available statements of ethical principles.

In addition, professions tend to focus on ethics not simply to address tangible immediate problems, but also to create a shared sense of occupational values. Under the framework of the international statement of ethical principles for social work (IFSW/IASSW 2018), the professional social work associations in at least 90 countries now either have well-developed codes of ethics or are in the process of creating such documents. The extent to which codes of ethics are established or are developing is a reflection of the length of time for which social work has existed as a profession. In addition, the creation of ethical codes has also tended to be a concern particularly noticeable in countries of the global North (Hugman 2013). However, the inclusion of global Southern perspectives has influenced more recent developments and sought to make the international social work ethical framework more inclusive (Sewpaul 2016). This framework explicitly calls on social work in each country to use it as a means to formulate ethical statements that are appropriate for their own cultures and societies, while remaining part of the global understanding of social work values in relation to its core goals and functions.

REFLECTIVE QUESTIONS/EXERCISES

What is the relationship between values and ethics in social work?

What are the key shared social work values or ethical principles in your country? What about the key shared values in your specific context of practice?

Virtue as a ground for professional ethics

For a long period of time the dominant perspectives that have formed the basis for social work ethics have been found in the principles of human dignity and worth, particularly expressed in human rights, and in social justice (Lundy 2006; Ife 2012). The international statement of ethical principles also emphasises the importance of integrity, which is understood in terms of the factors affecting the interactions of social workers with other people from the perspective of the ethical character and orientation of the social worker. In this way, social work ethics draws on the three classic traditions of deontology (duty to moral principle), consequentialism (in the form of social justice) and virtue ethics (moral character). However, McBeath and Webb (2002) argue that most of the debate about ethics in social work tends to focus on deontology and consequentialism, with relatively little attention paid to virtue ethics. This, they contend, diminishes social work's consideration of its values and can lead to a rigid, rule-based approach that removes the sense of the individual person within a social and cultural context that social work claims to recognise (compare with IFSW/IASSW 2018) and the necessary engagement of each social worker with moral responsibility and accountability.

Banks agrees with this criticism, noting that the principle-based ethics approaches do not address 'important features of the moral life and moral judgement, including the character, motives and emotions of the moral agent' (2012: 69). As the American moral philosopher Hinman puts it (2013: 46), virtue ethics addresses the agent rather than the action. This perspective leads Banks then to argue that although principle-based approaches can provide many ways of grasping values, they are not able to address the way that good or right actions are contextualised in 'the specific core purposes or service ideals of a profession' (2012: 74). That is, a virtue ethics is necessarily particular to time and place, specific, partial and, because it is related to personal qualities, different between different practitioners (2012: 70). For example, even though moral courage might be highly valued as a quality in practitioners, trying to specify this as a requirement is problematic because it will be expressed in various ways and to differing degrees depending on who, when, where and how it is present.

Understanding virtues in this way draws heavily on the tradition that is usually ascribed to the classical Greek philosopher Aristotle (384–322 BCE) (Hinman 2013; Papouli 2018). Such an approach regards moral character as closely related to the broader purposes of the social roles that a person performs. Non-moral virtues can even be ascribed to inanimate objects, such as when we say that the virtue of a clock is that it tells the time accurately or a knife that it is sharp (Pawar *et al.* 2017: 4). Similarly, non-moral values also apply to professions, such as manual dexterity in a surgeon, ability to communicate ideas in a teacher or the capacity to listen in a social worker. However, with human agents (here seen in the form of professional practitioners) non-moral values are less easy to disentangle from moral values in the case of people than inanimate objects. Medical practitioners exercise their practical virtues in the context of the pursuit of health as a moral value, teachers are engaged in the pursuit of learning and social workers in the pursuit of personal and social wellbeing (compare with Koehn 1994). From this, we may then say that the non-moral virtues of social workers lie in the technical qualities of their practice, such as whether or not someone has the capacity to conduct an interview, or make an accurate assessment of a person's situation, or assist people to advocate for structural change.

This understanding is supported by MacIntyre (1985: 154), who argues that some virtues can be taught through instruction while others need to be learned reflectively through practice. The virtues that can, indeed must, be taught he calls the 'intellectual virtues'. Those that can and must be learned through practice so that they become habits are the virtues of character, or moral virtues. However, there is a problem for professions such as social work in taking this distinction at face value. Some of the non-moral virtues of professional life, such as techniques, must be learned through both instruction and practice. Examples here include the qualities of being an effective counsellor, or advocate, or social development activist (compare with Pawar and Anscombe 2015). To address this point, we can consider the Aristotelian virtue of practical wisdom (or 'phronesis'). Instruction, which now takes the form of higher education in many countries (IASSW/IFSW 2004), not only helps the beginning social worker to learn the explanations for why certain practices can be considered good technically, but that through practice these become part of the character of the social worker. Then, as MacIntyre expresses it:

> [c]onversely the exercise of practical intelligence requires the virtues of character; otherwise it degenerates into or remains from the outset merely a certain cunning capacity for linking means to any ends rather than those ends which are genuine goods [as defined here by the profession].
>
> *(MacIntyre 1985: 154)*

An example of the way this occurs can be seen in the injunction in the international statement of ethical principles for social work that 'social workers should not allow their skills to be used for inhumane purposes, such as torture or terrorism' (IFSW/IASSW 2018: section 9.3). Despite being phrased as an imperative of moral duty, this statement appears in the section concerning 'professional integrity' and is intended as a concrete example of the qualities of respectfulness, justice and care, realised through the virtue of wisdom.

For this reason, non-moral virtues can also be said to be important for the moral values that a profession might claim. One example affecting many professions that attend to human needs is that of justice. Justice, understood as fairness or the balance between competing claims, is part of the ethical statements of allied health, medicine, nursing, social work and teaching (Hugman 2005a; Banks 2012; Beauchamp and Childress 2013; Strike and Soltis 2015). Moreover, each profession asserts this value both as a goal and a way of shaping practices, as well as a virtuous aspect of the character of practitioners (a point to which we return below). At the same time, paying attention to the technical skills of a profession (as non-moral virtues) cannot be separated from the goals of that profession, or from attention to questions of who gains access to them, as moral questions. Similarly, the use of other scarce resources (such as physical equipment, buildings and so on) combines both the non-moral idea of a virtue (the object is good for its purpose) and the moral question of who should benefit in a world where resources are always limited (a consequential ethic). Examples include the nurse making decisions about who should be attended to first in an emergency department or clinic, a teacher balancing attention to one student with complex learning needs and each of the others in a class of 30 students or a social worker considering who should be recommended for access to a specialist service for which demand exceeds availability. Although assessment of physical need will play a part in these deliberations (for instance, the nurse will prioritise someone who is bleeding seriously over someone with pain but whose life is not immediately threatened), psychological and social factors also are important in all such situations. Here questions that may be said to have a moral dimension will be addressed, for example, whether services should be made more available in a neighbourhood that is struggling economically than one that is wealthier.

Although the value of justice is often regarded as a question of principle, since Aristotle justice has also been seen as a moral quality of people. The just person is one who is inclined to seek fairness and equity and to oppose inequality and negative discrimination (Donaldson and Mayer 2014). The complexity of fairness across society is one that makes this a difficult value to pursue. Because of the scarcity of resources relative to need, and the many inequalities that some regard as 'entirely natural' (such as physical and intellectual abilities) or that come to be socially justified from some perspectives (such as economic inequality), what it is to be a just person is highly contested. For some practitioners it is found in the tendency to be impartial in dealing with service users, and for other practitioners it is seen in the tendency to prioritise those who are seen to be in particular types of disadvantage. We will return to the virtue of justice in Chapter 6.

The other major element of Aristotle's thought that is relevant to the discussion presented in the following chapters is that a virtue can be understood as the 'mean', or mid-point, between two extremes, which are referred to as vices (MacIntyre 1989: 62–63). In other words, to describe a quality as a virtue is to say that it represents an appropriate balance between too much and too little of that quality: it is not an arithmetic mean (Papouli 2018: 9). For example, if we take 'compassion' as an ethical virtue that is understood as an empathic response to the plight of another person, then the excess might be seen as 'pity' and the deficit as 'indifference'. (Compassion is discussed in depth in Chapter 4.) From an Aristotelian perspective, a virtue and its associated vices relate to what is good in the character of the person but also in the way that this quality contributes to the wellbeing of the wider society. This in turn raises the question of whether it is possible to be compassionate in a society that does not value compassion, which we address further later in this chapter.

The idea of virtue can also be seen in other cultural traditions. Confucian teaching, a major pillar of moral thought in Asia, stresses the responsibility of the good person as the foundation of a good society (Shun and Wong 2004). Some scholars are sceptical about the extent of the comparison that can be drawn between Confucius and the classical Greek philosophers (such as Socrates, Plato and Aristotle), whose ideas continue to influence global Northern thought (Lai 2006). However, there are some parallels between these philosophies in the focus on a person's moral characteristics as an approach to ethics, together with the determination of what is good and right being grounded in social relations and structures.

Similar approaches can be seen in African thought (Graham 2002) as well as that of Indigenous peoples of various parts of the world (Voss *et al.* 1999; Walker 2012; Weaver 2017). In some ways, these ethical approaches share the value of communality, as compared to the individualistic approach of European and other global Northern ethics. Mbiti's influential translation of the core meaning of 'ubuntu' into English is 'I am because we are; and since we are, therefore I am' (1990: 160). At the same time, within this communal ethics there is also a sense of what European and other global Northern approaches term ethical virtue. For example, Tutu (1999: 31) argues that 'ubuntu' is demonstrated through a person being 'open and available [to others]', a 'self-assurance' that comes from knowledge of being part of the community, combined with a form of respect that is shown by an awareness of being 'diminished when others are humiliated or diminished, when others are tortured or oppressed'. These are all personal qualities that contribute to a flourishing life, communal and individual, as much as courage, compassion or justice.

Likewise, the example of the value of 'yindyamarra winhanganha' in Wiradyuri culture (the Indigenous people of what is now known as central New South Wales, in Australia), translates as the wisdom of living respectfully in a world worth living in, gently taking one's time (Sullivan *et al.* 2016). In this culture, the values of wisdom, respect, care and gentleness are not separated, but encountered together in a single ethical virtue. The same complex rendering of values that European culture has separated can also be seen in New Zealand's Mäori culture and the various Indigenous cultures of North America (Voss *et al.* 1999; Weaver 1999, 2017; Walker 2012). For example, Weaver (1999: 222) notes that self-awareness, humility, respectfulness, open-mindedness and justice are interwoven at the core of Indigenous American values. Unlike the Aristotelian tradition, in which these virtues may or may not co-exist (and are not required to), for many Indigenous cultures it is unthinkable that a person could have one without the other in order to live well in the world ('flourish', in ancient Greek terms). As social work increasingly addresses issues of multicultural and decolonising theory and practice, and seeks to bring these processes to bear on the ethics and values of the profession, addressing ethical diversity within a shared framework has become ever more important, as has been recognised in the recent reviews of the international statement of ethical principles (Sewpaul 2016; IFSW/IASSW 2018).

More recent developments in virtue ethics in global Northern moral philosophy have broadened the scope of inquiry into moral character (MacIntyre 1985; Slote 1992, 2011; Hursthouse 1999). There is a shared theme in these contemporary approaches to virtue ethics. They argue for virtues as the important central question for ethics. That is, they each propose that attempts to ground ethics in deontology or consequentialism are inadequate and that moral character is at the heart of the way that people understand and act in the world. However, they then part company in how they achieve this. Hursthouse's (1999) argument is that an action is right if it is what a virtuous person would do. In some ways this is close to Aristotle's understanding. However, this approach leaves open the question of what is the character of the person who is acting, as the act is considered virtuous in terms of what someone else would do. That is, in what sense does this make it a virtue for the person acting, as opposed to following a moral rule that has been derived from rational thought? At the least, this runs the risk of being seen as deontology slipping in through the back door. However, it also recognises the importance of ethical virtue as learned, as well as the role of 'exemplars' in that process (that is, people who can be seen as ethical role models).

In contrast, MacIntyre (1985) proposes that ethical virtues can be seen in terms of their consequences in the actions of the virtuous person. That is, we understand the virtuous person to be someone who is inclined to think and do those things that will contribute to their own and the wider community's flourishing. There is a risk here that virtue becomes defined in consequences, but the source of moral authority remains the character of the person. Also, unlike Hursthouse, MacIntyre finds the meaning of virtue in the meanings that people give to their own lives, shared in families, communities and cultures. His approach is not to create principles about the actions that should be followed, which can be followed as if they were rules, but to encourage the development of good judgement as a central virtue. This then leads MacIntyre to say that virtues support good or right practices, which has something important to offer for a profession such as social work (Banks 2012: 72). While it opens the risk of ethical virtues being too relative to culture, at the same time it also creates the opportunity to avoid trying to identify virtues that would apply to *all* social workers in *all* situations, and perhaps to consider particular virtues in relation to roles. The virtues that we value in a counsellor might well differ from those valued in a social development practitioner.

Slote (1992, 2011) explains ethical virtues from yet a different point of view. His understanding of virtue ethics is one that brings together an emphasis on emotion and the insights of feminist 'ethics of care'. The meaning of emotion here points to the way in which we react to the world around us and it has a long history in what philosophers call 'moral sense theory'. This might also be referred to as 'having a moral compass'. Here too there is a risk of relativism making virtues too particular to be useful in thinking about ethics for a profession. However, that is avoided for Slote in the same way that it is for MacIntyre (1985), in that the definition of virtues comes from the shared sense of what is good and right among social workers – this sense is shaped and reshaped within a professional *community*. We can also note that understanding moral values through people's responses to their experiences can be seen, also, in the Confucian tradition (Lai 2006), which creates the possibility of intercultural ethical dialogue.

To summarise, in considering the moral character or quality of someone or something, we may refer to the notion of virtue. In this sense, virtue has two forms. First, it may refer to characteristics or qualities that define a person as good in relation to her or his function or purpose. Second, a different type of virtue can be identified, namely those that concern the moral dispositions that people may have, seen in relation to what is regarded as good and right in a particular cultural or social context. It is this last aspect that virtue ethics addresses and that is the focus of this book.

REFLECTIVE QUESTIONS/EXERCISES

Define 'virtue' in relation to social work ethics.

What are the differences between intellectual or non-moral virtues, and moral virtues? Why are both important in social work?

How would you define the 'just person'? What makes the value of justice difficult to pursue?

Virtues in social work ethics: taking a principled stand or a stand against principles?

So far we have identified both that professional ethics tends to be framed in terms of the principles used to evaluate goals and actions, and that while this is necessary it is not sufficient. Virtue ethics adds another necessary dimension, through its recognition of the moral character of the practitioner as a moral agent. However, when looking at the ethics of social work and other professions there is a tendency to examine the principle-based approaches, either predominantly or exclusively, and to pay less attention to the character-based contribution of virtue, or even none at all. To look in more depth at why this might be and how considering virtues might be included more explicitly in social work ethics, we now return to a key discussion of this area of professional ethics that is identified above.

McBeath and Webb (2002) present their critique of social work professional ethics based on a review of ethics literature, noting that in the last 50 years it has been almost entirely absent. Although since the original publication of this argument there has been a modest growth in attention to virtue ethics for social work (Houston 2003; Hugman 2005a; Pullen-Sansfaçon 2010; Banks 2012; Pullen-Sansfaçon and Cowden 2013; Papouli 2018), it remains the case that discussions of ethics in social work still focus very largely on principles, especially those of human rights and social justice. McBeath and Webb call this the 'incessant drone of Kantianism and utilitarianism' (2002: 1019). Even the ethical framework provided by the international social work organisations addresses virtues only in relation to the notion of 'integrity', in statements contained in one section (IFSW/IASSW 2018). In addition, although integrity is an ethical virtue (see Chapter 9), the ethics documents of the profession appear to understand it entirely in terms of specified 'conduct'. Furthermore, of the eight clauses in the international statement dealing with integrity, only three refer to underlying ethical ideas (respect for human dignity, responsibility and accountability).

Although all these statements are grounded in ethics, they implicitly draw on principles rather than qualities of the practitioner in such a way as to render virtues subordinate to principles, to say the least. (At worst, this may render any consideration of virtues superfluous.) McBeath and Webb (2002) argue that implicitly relying on principles in this way creates two problems for social work ethics. The first of these is that it turns what should be an engagement of practitioners in moral thought into a series of directions (somewhat like an instruction book or user manual). That is, it reduces what should be dynamic to being static (McBeath and Webb 2002: 1020). Life, they note, is contingent; it is impacted by factors of time, place and the actions of other people. This is what moral philosophers refer to as 'moral luck'; some people just have better circumstances in which it is easier to appear to 'act well', whether or not they do so, or actually to be seen to have acted well because their acts achieve the desired ethical consequences (McBeath and Webb 2002: 1023; compare with Nagel 1979).

Second, only relying on principles to guide ethics appears to support an expectation that ethics is a specialty, engaged in only by some social workers (who determine the regulations), and not the concern of non-specialists (other than to read and follow the manual). In contrast, McBeath and Webb argue that ethics involves 'play[ing] one's part' in defining and achieving what is good and what is right (2002: 1030). Ethics is everyone's concern and for everyone to be involved in realising. However, they also note (2002: 1032) that the more situational, fluid approach to ethics that this would produce brings with it less scope for organisations to seek to control risks through controlling the actions of practitioners.

Banks' review of virtue ethics in social work (2012: 71–77) notes that in practice, agreeing on the key virtues for social workers has proved rather difficult. In a review of four ethical documents that include the notion of virtue Banks (2012: 74) identifies only one, that of 'moral courage', as a commonly claimed ethical virtue. She also notes that almost all of the virtues that are asserted in this sample of statements are quite general and could equally be applied to most other human service professions, if not to most people living their everyday lives. This in turn leads Banks to suggest that a virtue ethics approach must be spelled out in relation to the central purposes or ideals that inform social work practice. However, despite then using this approach to offer a particular list of ethical virtues for social work (which we will consider further in Chapter 2), Banks concludes that even producing these ideas does not resolve the issue of how to define these virtues, or even how to separate them from principles, noting: '[i]t could also be argued that virtue ethics can be subsumed within principle-based ethics; that being a just person simply consists of acting justly' (2012: 76).

Houston (2003) identifies that all of these questions present a clear challenge to the way in which virtue ethics can be considered (in this case, for social work). Trying to find ways of defining virtues is the central problem. Most discussions of virtue ethics still draw on the Aristotelian tradition, but the contemporary world cannot rely on a shared set of cultural values as the basis for virtues, as was the case for Aristotle in ancient Greece. Furthermore, Houston argues (2003: 820) there is a circularity in the Aristotelian approach to virtues that underpins most such discussions. Looking at the character traits of virtuous people to see what qualities they possess assumes the prior definition of the virtues for us to be able to say who are virtuous people. Trying to overcome this limitation by using the classic Greek notion of 'a flourishing life' (a life lived well) to define a virtuous person does little to resolve the matter, as the definition of what it is to 'flourish' in this sense itself includes living a virtuous life. This in turn might be circumvented by defining ethical virtues in terms of what produces wellbeing for the self and all those around in the community. However, this is bringing in a consequential measure that explains character in terms of actions. Virtuous character is the disposition to act in certain ways, but it is not the act in itself that is the measure of rightness. Then, relying on moral intuition to tell us what the virtues are is also prey to relativism, as each person may value different personal qualities; thus there are risks also of personal bias to cultural bias (Houston 2003: 820). Finally, Houston asks if there is a compelling argument against abandoning virtues along with principles and instead following Nietzsche in the pursuit of power, or Kierkegaard in existential freedom from frameworks altogether (2003: 821).

Nevertheless, Houston clearly thinks it is worth social work finding an alternative to Nietzschean nihilism or existential individualism. The solution he offers (2003: 821–823) is for social work to seek a shared, reflexive and interpretive process by which an understanding of professional (ethical) virtues and the virtuous practitioner might be insights that are developed out of an interactive dialogue in which all voices can be heard. He attempts to deal with the question of differential access to social power by ensuring that the conditions in which an ethics is formulated are freely open to all (Houston 2003: 822). In one sense, in many countries the profession itself usually manages the development of its ethics. In those societies, the relevant professional bodies produce their ethical documents. In such circumstances we could say that the conditions described by Houston might potentially be met. At the same time, in this approach it is also important to consider how voices from outside the profession itself can be heard, especially those of service users, of the other professions with which social workers work, and of the wider society. In those countries where professions are not independent of the state in this way, the opportunities for achieving this are much more constrained. Houston does not address the contributions of twentieth-century virtue ethics scholars such as Hursthouse, MacIntyre and Slote, but, as noted above, they emphasise this sort of communal, dialogic approach to the problem of circularity in defining ethical virtues.

There have been some attempts to use research as a way to access the voices of service users that have been widely publicised across social work, including some which produced insights that have now passed into 'taken-for-granted' professional wisdom. Examples of these include Truax and Carkhuff's (1967) groundbreaking study that provided evidence from practice for warmth, genuineness and empathy as key qualities of practitioners. Others that followed include Mayer and Timms (1970), Sainsbury (1975), Rees and Wallace (1982) and Sainsbury, Nixon and Phillips (1982), who variously identify honesty, fairness, promise-keeping, respectfulness and trust (or trustworthiness) as important qualities of social workers, among others. While it has become more prevalent in research to include service users as sources of information about social work practice, as well as having service user voices in the planning and provision of some social work services (for example, Kvarnstrom *et al.* 2012; Du Plooy *et al.* 2014), these earlier studies were more explicit in their attention to the views of those receiving social work concerning the qualities of practitioners. At the same time, there remains very little systematic evidence about the way in which members of other professions perceive the qualities of social workers. Beyond this, views held more widely in the community appear often to be affected by political convictions about the wider purposes of social services (for example, Briar-Lawson *et al.* 2011; Warner 2014; Beddoe *et al.* 2019). So although there is some way in which social work might engage in the approach suggested by Houston (2003), the processes for doing so are not part of most professional structures and require careful thought.

Banks (2012: 76) also provides a more prosaic caution about virtue ethics as a framework for a profession, namely that by its very nature a virtue approach to ethics does not easily lend itself to codification. As we have already noted, discussions of virtue ethics make clear that virtues are contingent on the character of the practitioner, on the purposes of practices, on their settings, and so on. By definition, a code or statement of principles has to be general and to cover all social workers, whether pitched at an organisational, national or international level. It is difficult, perhaps even impossible, to make definitive statements about ethical characteristics of practitioners that can be held to apply to all practitioners in all circumstances. There is an argument that, although virtue ethics cannot be specified in that way, it is at the same time possible to devise some principles from virtue ethics that can be spelled out in relation to particular settings or consequences, such as the goals of social work that are justified from a shared understanding of the good society (Papouli 2018).

Does this then mean that virtue ethics cannot form an appropriate focus for a profession such as social work, or at least is limited in how it may do so? While it may not be possible to say with certainty that the foundation of an interactive dialogue as proposed by Houston (2003) exists, at the same time it is also not plausible to suggest that there is no opportunity to consider virtue ethics or that there is no value in doing so. Much depends on the purpose and scope of such an exercise. If the goal is seen as replacing principle-based ethical frameworks, this is unlikely to be supported when at many levels social workers are highly committed to the defence of human rights and the promotion of social justice (O'Brien 2011; Ife 2012; Reichert 2017; Reisch 2017). However, if the intention is to make virtue ethics more intelligible for social workers and to promote wider consideration of virtue ethics as a basis for understanding the complexities of a consciously ethical practice, then this approach has great potential to connect abstract ideas to the particularities of everyday practices.

We should also be clear that the goal of providing an exclusive alternative to principle-based approaches is not the intention of the present authors, whose interest is instead to add to the richness of a plural approach in social work ethics (Hugman 2005a, 2013; Pawar and Anscombe 2015; Pawar *et al.* 2017). Rather, we assert, there is much that can be promising for the profession in social workers seeking insights about ethical virtues in social work practice alongside attention to principles of human dignity and worth, human rights and social justice (compare with Houston 2011). This is the way that we might move closer to achieving Houston's (2003) proposed critically reflective dialogue.

We have to sound a note of caution here, as we already observed above that Aristotle's approach to virtues assumed a particular social and cultural context. Although we have questioned the plausibility of this in a multicultural world in which we find value pluralism, we are not advocating that all aspects of this idea should be abandoned, even though it presents potential problems of relativism or of uncritically accepting the status quo. What we are advocating is that rather than being able to make definitive positive statements of the social and cultural values that are required for a characteristic to be regarded as a virtue, what is possible is to consider those influential values that impede a characteristic being realised as a virtue. To return to our previous example of justice, the person who is just will have great difficulty in being seen as virtuous by those who dominate in an unjust society. A similar fate awaits the compassionate person in a society where indifference predominates, or the person of integrity in circumstances where 'by whatever means' is the overriding value. It is for this reason that in some geopolitical circumstances social work can be regarded as troublesome, because the ethical virtues of the profession, together with other values (dignity, rights, justice), may be claimed in stark contrast to the dominant surrounding values. For social work, ethical virtues, as much as principles, are interwoven with the profession's identity, asserting *these* values and that practitioners should develop *these* qualities. We return to this plural construction in the next chapter, drawing on the definition of social work that is widely used to inform the professions ethics globally.

REFLECTIVE QUESTIONS/EXERCISES

What are some of the limitations of principle-based approaches to ethics?
How might virtue ethics address some of these limitations?
From the overview of existing literature above, what are some of the key virtues in social work that you might like to explore further?

The basis of this book and its structure

By beginning with the arguments presented by McBeath and Webb (2002), Houston (2003) and Banks (2012) as a basis, we can start to identify ethical virtues from the views of social workers and others that are available from a range of sources. Yet our purpose is more than only analysing these discussions from a philosophical perspective (useful though that may be). It is also to set out some arguments for a particular view of ethical virtues and how these may be considered for social work. Our discussion draws on a number of sources, which complement each other to provide a variety of viewpoints.

First, our major inspiration for this present discussion comes from the ideas shared with us by social workers in Australia as part of an Australian Research Council Discovery Grant-supported study into virtue ethics in the profession. The project had three phases, the first of which used a biographical approach to examine the life and work of ten 'exemplars'. That phase, which has been published (Pawar *et al.* 2017), suggested a range of characteristics of 'good social workers' that is reviewed in more detail in the following chapter. The second phase involved interviews with 65 social work practitioners and supervisors around the country, with coverage in each State and Territory. The methodology used was based on open-ended qualitative interviews conducted by some members of the team, which were then analysed inductively by other members of the team. The resulting themes are also summarised in the next chapter and quotes from the interviews are used throughout this book to illustrate the argument that is developed. While the book is not strictly a research report, it is research-informed in this sense. The third phase of the project was a review of the ethics content of qualifying degree programs in Australian universities and other higher-education institutions (Pawar *et al.* 2019). This phase was partly pre-coded, with some element also of qualitative responses. Thus the analysis used a mixed-methods approach and, although the resulting information is quite descriptive, it also reveals some details about the extent of attention to virtue ethics in foundational professional social work education.

The second source of ideas in this present discussion is drawn from literature from around the world. Notwithstanding its origins, this is not a book only about Australian social workers, but about social workers in many contexts. We have already noted the question of particularity in considerations of ethical virtues and, while we want to avoid slippage into a fixed set of qualities that should be applied equally to all social workers everywhere, it is also our aim to address the diversity of the profession. The literature that addresses ethical virtues in social work is growing, if slowly, and there are contributions emerging from various parts of the world. Using this material enables us to broaden the discussion from our own base in Australia to consider other, different perspectives.

Then, third, four of the team are qualified and experienced social work practitioners, of whom two are involved in different ways in international social work; the other team member is a philosopher who can ask critical questions from the 'outside'. Our own experiences, observations and knowledge thus provide another point of reference, enabling us to look both at the ideas developed from critical reflections on our own perspectives as well as those of the many others from whose ideas we have learned.

This chapter has set out the background issues of virtue ethics in social work. It identifies the problematic place of virtue ethics in the broader ethical conversation of social work and in professional ethics generally. Chapter 2 looks in greater depth at the debates about the place of 'virtues' in developing good practice and the challenges for engaging with virtue ethics in social work compared to some other professions. In particular, the attention to virtue ethics in medicine, nursing, allied health and teaching, is used as a comparison from which to consider both why and how virtues should be given greater attention by social workers. The chapter presents a summary of the virtues that form the structure of Chapters 3 to 9 and provides further explanation as to how these were arrived at. In conclusion, it argues that virtue ethics applies to all modes of practice, including micro, meso and macro practices. It may even challenge the rigidity of these distinctions (McBeath and Webb 2002: 1033).

Chapters 3 to 9 each examine a particular ethical virtue or cluster of closely related virtues. In Chapter 3 we consider 'courage' (which, as we have noted above, appears strongly in many discussions of virtuous qualities in social workers). Some commentators even consider courage to be the key virtue, and we consider this argument.

Chapter 4 looks at compassion and care, examining not only the way that these two virtues relate to each other, but also the way in which they provide a ground for the quality of 'empathy', which is widely regarded as critical for good social work practice. This chapter also carefully explains some commonly confused ideas, for example, distinguishing between compassion and pity and differentiating care as a virtue from the ethical approach of 'the ethics of care'.

In Chapter 5 we consider the related virtues of hope, resilience and perseverance. These are examined as qualities that enable social workers to sustain their commitment to long-term, entrenched and even (apparently) intractable issues to which they respond. Humour is also addressed as a possible contributing quality in this regard.

Chapter 6 examines justice as an ethical virtue of social workers. It differentiates justice as a virtue from justice as a (consequential) principle, to enable the discussion to explore the relationship between these two values as well as to look at the ethical virtue of justice as a basis for various practices. In this way, justice is examined in the relationships between social workers and service users as well as between service users and the wider society.

In Chapter 7 we look at humility as an ethical virtue. The discussion presents an understanding of humility, based on a balanced view of the self in relation to others, and explores the way in which humility may contribute to the capacity of practitioners to recognise and respond to 'the other'. In particular, this discussion also addresses the meaning of cultural humility and the way in which social work engages with social and cultural difference, not only in relation to ethnicity but also to other forms of social structural division and identity. This chapter also considers the challenge of the exercise of power in practice for humility and spells out a way of addressing the apparent contradiction that provides social workers with reconciling the competing ethical demands of power and humility.

Chapter 8 focuses on wisdom, asking what it means to think about this as an ethical virtue in practice. The chapter presents an explanation of wisdom as the capacity to combine thought and action across a range of practices. It then examines the challenges faced by social workers in applying practical wisdom in the exercise of judgement in contexts where increasingly goals and methods are prescribed. From this, the chapter offers an argument that the development of practice wisdom is a vital part of ongoing professional growth, as a quality that can be learned and developed with conscious attention.

Then, in Chapter 9, we address integrity as the ethical quality of integration. In one sense integrity can be seen as the virtue of connecting other virtues in the character of the practitioner, such as in achieving a balance of honesty, consistency, trustworthiness, respectfulness and temperance. The place of each of these virtues is discussed to explore how integrity is such an overarching virtue. The chapter then addresses the way in which integrity also plays an important role in the connection between the virtues that are examined in the preceding chapters.

Finally, in Chapter 10, we summarise key points from the other chapters and address the overarching question of the ways in which each of the virtues discussed in Chapters 3 to 9 has something to say about ideas of 'the good social worker' and good practice. The central argument is that each of these virtues has relevance for both the profession and individual practitioners. This chapter also looks at the ways in which different virtues are emphasised differently between practitioners, fields of practice and even specific instances of practice, such that there is no single model of one type of person who is suited to social work. Instead, the argument that is offered suggests a way of understanding and nurturing virtues for the realities of practice. In conclusion, the chapter argues that virtues must be considered alongside human dignity, human rights and social justice as central to social work ethics and as a way of helping social workers to strengthen their practice. The final point presents an argument that although the cultivation of virtues is a responsibility of each social worker, at the same time the profession as a whole must attend more explicitly to this element of ethics.

Throughout the chapters, examples from practice taken from our own research and from existing literature are used to strengthen this discussion. These examples are used illustratively, in the traditions of both virtue ethics discussions and also qualitative research, and not as definitive 'proofs'. As we have argued in this chapter, ethical virtues connect moral qualities or character with practice, so these illustrations provide the clear means to ground and extend our understanding.

FINAL REFLECTIVE QUESTIONS/EXERCISES

List three key learnings from this chapter.

The authors argue that virtue ethics has not been attended to very explicitly by the social work profession. Why do you think this is the case?

Think of a 'good social worker' you know – someone you would consider a moral exemplar. What are the qualities that stand out to you? How/what have you learned about social work from this person?

2

VIRTUE (ETHICS) IN SOCIAL WORK PRACTICE

CHAPTER OBJECTIVES

- To outline the ethical virtues selected and examined in this book
- To explore the contributions of business, health and education professions to social work's understanding of virtue ethics
- To describe the relationships between different moral qualities and how these connect in social work practice

Which virtues?

In the previous chapter we examined the ideas of professional ethics and virtue ethics. The discussion highlighted that although virtue ethics has not played as important a role in framing social work ethics as have the principle-based approaches of deontology and Utilitarianism, virtue ethics has much to offer the profession. In this chapter we consider the ways in which virtue ethics has been addressed in the human service professions and then go on to look at the evidence from the limited range of studies that have focused on social work.

Part of the agenda here is to establish the ground for Chapters 3 to 9, in which we consider in depth particular ethical virtues that can be identified as important for social workers. To do this, we draw both on the key literature, which addresses social work in several countries, as well as ideas that have been obtained from research and professional experience in Australia. The voices of practitioners are central to this discussion and many of the insights that are shared throughout this book reflect the experiences of social workers in a wide range of fields of practice. The ethical virtues that are examined, therefore, are those which embody

MacIntyre's (1985) view that, whether we base our understanding of virtues on Aristotle's model or that of rival approaches, the (ethical) virtues are always embedded in and reflective of practices (van Hooft 2014).

Learning from other professions

Previous attention to questions about the specific virtues that can be identified for social work has made comparisons with the wider application of this approach in other professions, especially including those that are concerned with health and human services, but also including other professional areas in which there is an identifiable practice (Pullen-Sansfaçon 2010; Banks 2012; Papouli 2018). These three discussions between them refer to allied health, business, medicine, nursing and teaching. For the purpose of establishing the common ground from which virtue ethics in social work can more consciously be examined, it is worth reviewing these comparisons, in order to identify possibilities and problems in thinking about ethical virtues in professional practice. Three broad professional areas are used to structure this review: business, health and teaching.

Business ethics may not appear to be the most obvious comparator for social workers, yet there are some important aspects that can assist our thinking. Solomon (1992), in an influential analysis in this field, notes that ethics in business ought not to be regarded any differently than ethics in everyday life. The moral values that Solomon emphasises in such a position are honesty, fairness, trust and fortitude (he uses the term 'toughness', but defines it carefully to make clear that he regards brutality or callousness as its vices of excess) (1992: 207–215). Subsequently Solomon adds friendliness, honour, loyalty and shame as qualities of the good businessperson (1992: 217 and following). In particular, according to Morse (1999), Solomon's position rebuts arguments that claim business is not to be evaluated from the same morality as other areas of life. Examples include the view that unlike family and community relationships, or indeed other professional spheres, business is like poker – a tough game in which adherence to the rules is the only moral value – a variation of which is that not breaking the law is the only moral requirement (Morse 1999: 48–49). Morse is particularly critical of the view associated with the economist Milton Friedman, that in business values are set by the goal of the actions involved, so that the only responsibility is to make a profit and there is no wider moral concern (in Morse 1999: 49). In such a perspective, it makes no sense for business to be concerned with social justice or civic duty. Against this, Morse argues strongly that a business is a community, and business takes place in and depends on a wider community; consequently, everyday values cannot be superseded in this way. Similarly, Morales-Sánchez and Cabello-Medina (2013: 718) assert that business is subject to the same moral competencies as is every other aspect of social life, namely sensitivity, judgement and motivation, which are integrated by character (by which they mean the capacity to connect the other three elements into action). Without this integration, business becomes divorced from the very society within which it has any meaning beyond the material accumulation of objects; wealth is a social and cultural concept.

There are two ways in which business ethics raises questions for social work. First, the prevailing approach to ethics that Morse criticises focuses on function as the basis for values – in this context, it is the function of business to make a profit. Social workers also perform socially mandated functions. However, as with ethically good business practice, it is not sufficient that social work values are reduced to the bureaucratic procedures of their service organisation (the equivalent in business to avoiding moral responsibility). Second, as with the business practitioner, social workers may reasonably be expected by the wider society to take moral responsibility individually as well as collectively. As Bauman (1994: 13) puts it, 'Bureaucracy strangles or criminalises moral impulses, while business merely pushes them aside'. Yet neither position is tenable. Moreover, it may confirm what many social workers have long suspected, that the neo-liberalism that is encountered as an obstacle to social work values has its roots in what appears to be an amoral approach to the relationship between means and ends and a refusal to accept that there is, indeed, such a thing as society. In the present context this refusal can also be seen as one that denies the relationship between society and the natural environment, the interdependence between them and the ethical challenges that arise from this (Dominelli 2012; Alston 2013).

Compared to social work, considerable attention has been paid to virtue ethics in the health field. Oakley and Cocking (2012) specifically address medicine in their influential discussion of virtues and professional roles, comparing medical doctors to lawyers. Notwithstanding isolated references to a particular (very specialised) branch of nursing (2001: 140, 146), Oakley and Cocking might be criticised for relying on very high status and historically masculine professions as representational of 'professions'. This avoids the explicit question of whether particular virtues may reflect social divisions of class, gender, ethnicity and so on. However, a major theme in their analysis is that of the struggle between human relationships as the basis of professional practices and arguments for detachment of the professional person from the beneficiaries of their work. Oakley and Cocking respond to this point by arguing that for the virtuous practitioner there is a limit to professional detachment. For example, without a degree of engagement with service users it would not be possible to achieve empathy (2001: 149). In other words, it is not possible to be a good practitioner without engaging emotionally and psychologically with the beneficiaries of practice in some way, unless such detachment serves the overall goals of the profession and that such goals are to 'serve morally worthwhile ends' (Oakley and Cocking 2001: 149).

In a previous discussion Oakley (1992) argues that emotions are part of a range of personal characteristics that may be used to evaluate a person ethically. In their view that detachment acts against more integration for members of professions, Oakley and Cocking (2001) are offering the one clear statement of 'virtue' in their analysis, that of integrity (when they otherwise refuse to name particular virtues). There are some parallels here with debates in social work, around the elements of good practice noted in Chapter 1, of empathy and 'genuineness' (Truax and Carkhuff 1967), or of the notion of 'controlled emotional involvement' (Biestek

1957), in which the importance of commitment has to be balanced with sufficient care of self to be able to keep going, with an implicit sense of Aristotelian balance. Yet, the key to these ideas is that emotional involvement and commitment are necessary for the good social worker: this cannot be achieved without 'caring about' both the social issues that social workers respond to and 'caring for' the people who experience those issues (Hugman 1991). Banks (2012) is correct to note that, like 'justice', 'care' is both a virtue and an underlying ethical value. In this sense the approach of care ethics (for example, Tronto 1993) is not a virtue ethics. However, from a pluralist perspective it must be seen as having much to say to an analysis of virtue ethics, precisely because it argues against the separation of emotion from human relationships in professional life. Recognising that emotions and commitments are central to who we are, and therefore to the virtues that we might seek to cultivate, is an insight that has been emphasised by a more conscious consideration of diversity in social life, on gender, ethnic or other grounds (Sevenhuijsen 1998; Kittay 1999; Nussbaum 2001; Clifford and Burke 2009; Bozalek 2013).

Other discussions of virtue ethics in medicine have been more willing to identify specific virtues. For example, Pellegrino has made a sustained contribution to virtue ethics in medicine (Pellegrino 2005, 2007; Pellegrino and Thomasma 1993). Pellegrino was himself a medical practitioner and sought to ground medical virtues in practice. The range of virtues that his work identifies as relevant to good medical practice are trustworthiness, compassion, practical wisdom, justice, fortitude, temperance, integrity and self-effacement (Pellegrino and Thomasma 1993). An overarching virtue, which is a consistent theme of Pellegrino's approach, is that the good medical practitioner places the patient at the centre of health care, as a guiding ethical commitment. Here it is termed 'self-effacement', which can also be considered a form of humility. This is an unusual way of approaching this virtue, and was not often seen before this in medical ethics. It embraces altruism in the sense of not putting the self first, and the lack of it in selfishness is seen to lead to what Pellegrino and Thomasma call 'the moral malaise of the professions' (1993: 145). This malaise can be seen in issues of incompetence, fraud, mismanagement of resources, sexual and financial impropriety, over-servicing of those who can pay and 'dumping' of those who cannot and so on. In that sense, self-effacement also embraces integrity, and contributes to the organisation of other virtues.

Similarly, the highly influential medical ethicists Beauchamp and Childress (2013) seek to identify key virtues for practice, at least in more recent editions of their key text. Where their work differs from that of those whose main focus is on virtues, is in the inclusion of these virtues alongside values of respect for autonomy (drawn from deontology), justice (drawn from Utilitarianism), beneficence (seeking to do that which is good) and non-maleficence (seeking to avoid doing harm), in the form of principles that are expressed deontologically as obligations. Nevertheless, they advocate virtues as the organising moral character of practitioners, which enables ethical decision-making in complex situations. For Beauchamp and Childress the virtues in medical practice are compassion, discernment, trustworthiness, integrity and conscientiousness (2013: 37–44).

Although they place moral virtues in a pluralistic framework with the four principles that they argue are the ethical foundation of biomedical practices, Beauchamp and Childress also allow that virtue ethics raises a question about the limits to moral responsibility. They note (2013: 46–47) that a person cannot be required to go beyond the demands of ethical duty, seen from the perspective of principles. The ethical notion of 'supererogation' in principle-based approaches refers to 'going above and beyond [moral duty]', whereas in terms of ethical virtues such qualities can be held in degrees of excellence. So, even when it is feasible to define obligations in the sense of a minimum ethical requirement, it is also possible to respect or even admire someone who seeks to go further, for example, by asking how someone else can be maximally treated well and minimally harmed (Beauchamp and Childress 2013: 51).

With regard to nursing, Armstrong looks for the ground of nurses' virtues in their roles and functions, but immediately finds that the profession is too broad to make this feasible (2007: 16–17). The conclusion, in very short summary, is that the good nurse is one who cares for people with ill health, combining moral qualities with practical skills. By implication, without the moral qualities the nurse who is technically skilful may still not be a good nurse, because the connection between means and ends is not realised. Yet Armstrong is reluctant to specify central moral qualities in the good nurse (2007: 31–32). He states that any such list, while seeking to be definitive, will inevitably be partial according to the perspective of the person who is compiling it. Implicitly, Armstrong also advocates for the virtue of integrity as central to understanding what it is to be a good nurse. Similarly, although Newham (2015) argues that by placing a focus on practice it may be possible tentatively to suggest particular virtues that are seen in particular situations in nursing, the only virtue that is named is that of *phronesis* (practical wisdom). Again, the virtue of integrity is assumed, but Newham (2015: 49) goes further than Armstrong (2007) and concludes that while virtues are necessary in being able to say what it is to be a good nurse, they are not sufficient. Thus there is a beginning of a pluralist account of nursing ethics, which is an argument that we reconsider in Chapter 10 of this present volume. However, both Armstrong and Newham are more concerned with considering the relevance of virtue ethics for nursing than being able to specify particular virtues.

Unlike Armstrong and Newham, Arries (2005: 69) is more definite and identifies particular virtues in nursing, including care, respect, integrity, justice, courage, wisdom, honesty and trust(worthiness). Yet, like Newham (2015), Arries also seeks to place virtues within a pluralist framework:

> [...] it would be inappropriate for nurses to act courageously if there was nothing worth acting courageously about, for example in situations where no facts are available or accusations that are purely based on hear-say [*sic*] [...] the acts of a courageous nurse are relevant to the particular circumstances of a case.
>
> *(2005: 67)*

So, for this reason the virtue of practical wisdom plays a part, not only in being able to make ethical judgement about which situations require a courageous response, but prior to that in knowing how to make judgements between the principles that might guide that response in relation to situations (rights, justice, beneficence, non-maleficence and so on). Social work practice, in this sense, is very like nursing in that it too may require the practitioner to exercise courage in defending or advocating for those who are already judged by others, while having the wisdom not to discern between claims for assistance. Thus, both the range of ethical virtues offered by Arries (2005) and the issues in understanding them in relation to practice have something to offer our present concern with virtue ethics in social work.

In the profession of teaching, as in social work, until recently there has been less attention to virtue ethics (Carr 2007; Cooke and Carr 2014). Cooke and Carr (2014: 91) attribute this to an over-emphasis on functional or technical competence. They argue that while this is something to be valued in any social practice, it is not sufficient to use as a basis for evaluating the goodness of that practice. In this view they also draw on Carr's earlier argument (2007: 369–370) that the moral character of practitioners in itself also matters for the impact that their practice has on those who they are there to serve (namely, their students). While it is recognised that ethical virtue in itself is not sufficient to enable students to learn particular subjects, nor is the capacity to structure a class or knowledge about that subject sufficient without virtuous character (Cooke and Carr 2014). Thus, from this perspective, the core virtues of teaching are seen as honesty, fairness, respect, patience and compassion, together realised through practical wisdom (*phronesis*).

Culham and Bai (2011) connect virtue ethics and emotional intelligence in professional teaching. Although they make reference to some of the moral virtues identified by Cooke and Carr (2014), their greater concern is to examine how the development of the appropriate virtues calls both for membership of a community, with traditions and shared values, and emotional intelligence (Culham and Bai 2011: 40). Their account of emotional intelligence has strong parallels with the more Aristotelian idea of practical wisdom, in that this virtue is both moral and technical, representing the capacity to make good judgements that integrate a person's own emotional and psychological responses to the world with the values of good practice, in the professional context. Teaching, as the professional practice in education, necessarily engages with the emotions as well as the cognition of teacher and student in a relationship. Here too we can see parallels with the moral 'goods' in social work that are identified above, of genuineness and controlled emotional involvement. Like them, the importance of emotional intelligence connected with ethical virtues is that it provides a means to those values as ends to be achieved in practice.

REFLECTIVE QUESTIONS/EXERCISES

What are the key contributions of the professions discussed above (business, health care [medicine and nursing] and education) to our understanding of virtue ethics?

List the key virtues as noted by the authors representing each profession.

Taken together, the core arguments for a virtue ethics approach to professional practice contain several common threads. These are that ethical (or moral) virtues:

- are located in a community of the profession, which is itself set within the wider society;
- are mediated by emotion and personality as well as thought, so their development requires attention in all aspects of the person;
- are therefore grounded in the purposes (or functions) that give meaning to the professional practice;
- may variously include such qualities as care, respect, integrity, justice, courage, wisdom, commitment (or fortitude), balance (or temperance), patience, honesty, trustworthiness, friendliness, honour, shame and loyalty, but that there is no agreement even in relation to a specific profession that a given group of virtues will be agreed to by everyone.

We can also see in this review of other professional fields that there are some virtues that mediate or combine others for practice. These include practical wisdom and integrity, which are also in themselves ethical virtues. There are also virtues that can be understood more as 'technical' virtues, which while not moral ends in themselves enable the moral virtues to be accomplished. These are emotional intelligence, controlled emotional involvement and genuineness.

We began this review of other professional fields by noting that others who are concerned with social work had pointed in this direction without themselves looking at it in depth (such as Pullen-Sansfaçon 2010; Banks 2012; Papouli 2018). In order to use these arguments as the starting point for our consideration of social work, we need now to turn to a more detailed examination of the way in which these questions affect social workers, as well as the distinctive contribution that social work can make to thinking about virtue ethics.

REFLECTIVE QUESTIONS/EXERCISES

What are the virtues that mediate or combine others in practice? How and where do you think these virtues are seen in the practice of social work?

Define 'controlled emotional involvement'. How might virtue ethics inform the balance between emotional detachment and engagement with service users?

Framing social work virtues

An increasing number of social work ethics discussions refer to virtue ethics, such as the more general analyses of Beckett and Maynard (2005), Hugman (2005b), Bowles *et al.* (2006), Parrott (2010), Bibus (2013), Bisman (2014) and Reamer (2018), each of which treats virtue ethics as part only of a much larger set of concerns, or even in some instances as secondary to the predominance of principle-based ethics. A smaller number have focused more specifically on virtue ethics in the form of arguments about how virtue ethics may be applied to social work as a primary focus, and we concentrate on these here: Webb (2010), Banks and Gallagher (2009), Pullen-Sansfaçon (2010), Banks (2012) and Papouli (2018).

First, these key discussions of virtue ethics in social work all in some way draw on the Aristotelian tradition of virtue ethics. In particular, at their core they concern the moral character or personal qualities shown by individual people. Although these characteristics or qualities gain their meaning from the context in which they are seen, these are not abstract intellectualisations that a person can employ as a rationalisation ('what is my moral duty?') or for making decisions in the form of a calculation ('what is the greatest balance of wellbeing to be achieved here?'). Virtues in professional life concern the moral qualities that are brought to the helping relationship by the practitioner.

Second, also originally from Aristotle, they address two other related elements. They consider that, to some extent, we can understand the moral virtues as those that help us to achieve 'human flourishing'. In other words, these are the qualities that are helpful to people living a life that is worth living. For a profession such as social work, understanding human flourishing has to include not only the well-being of the person acting but also, perhaps even more so, the person or people who are the beneficiaries of the act (service users, whether individual or collective) (MacIntyre 1985: 148). As we have noted above, Aristotle also thought of virtues as being embedded in a community and not simply as expressions of individual preference (MacIntyre 1985: 150; compare with Houston 2003). Taken together, these two factors mean that although ethical virtues are to be seen in each person, they are not to be regarded as 'individualistic' in the sense that we are each free to choose which virtues we happen to find congenial solely because of the value that they hold for us personally. They might have this sort of significant meaning for an individual, but that is not what makes a particular set of qualities a moral virtue.

Third, social work discussions of ethical virtues recognise that this approach concerns not only a disposition in people but also, necessarily, the way in which dispositions are integral to practices. For MacIntyre (1985: 187 and following) this follows from the foundation of the virtues in an ethical community. As Banks and Gallagher observe, MacIntyre's examples of practices tend to focus on such activities as playing music or sport. However, they note:

> [h]e does not dwell on professions as practices, but clearly MacIntyre's conception of a practice fits with the concept of a profession as a community of practitioners undertaking activities that are directed towards particular goods and ends.
>
> *(Banks and Gallagher 2009: 46)*

This is a helpful point, as MacIntyre defines practice as:

> any coherent and complex form of socially established human activity through which goods internal to that form of activity are realized in the course of trying to achieve those standards of excellence which are appropriate to, and partially definitive of, that form of activity.
>
> *(1985: 187)*

In other words, 'social work practices' are a regular pattern of what social workers do that is distinctive to social work and expresses the values of the profession.

Banks and Gallagher continue (2009: 47) by adding that practices are not the same thing as institutions (hospitals, clinics, social work departments, community centres and the like). Practices may occur within institutions, but it is important to distinguish them in order to understand their relationship to each other (which, as we see in later chapters of this current volume, helps to explain why ethical virtue is sometimes problematic for institutions). In this sense, practices are what people do.

Fourth, the way in which there is a lack of agreement across any profession about the definitive list of virtues that might apply might be considered a strength rather than a weakness in the argument for a virtue ethics approach. All the professions considered here, whether business, medicine, nursing, teaching or social work, consist not simply of one practice but of a range of practices, at least in the sense that skills and knowledge are related to a range of different needs and objectives. The virtues required by a psychiatrist may be different from those of a surgeon, the virtues of a community nurse may differ from those of a nurse in the emergency room, or those of a primary school teacher from a teacher in senior high school, even where there are some common threads. Aristotle is credited with the observation that playing the flute well is the moral goal of the flute player, and that the virtues of the flautist do not have to be comparable to those of the violinist (Sandel

2011). For this reason it may be desirable that people with different dispositions are members of the same broad profession, while focused on quite different forms of practice. In social work, the good counsellor may well have different qualities from the colleague who administers a long-term care program, and the good community development practitioner can be expected to differ again.

These points pose a crucial question for social work, namely whether the distinctions between modes of practice (micro, meso, macro) lead to the possibility that those are more important ethically than what unites social workers under one overarching notion of the profession. In other words, does virtue ethics raise the possibility of there being several 'social works' as opposed to a single profession of social work? One way in which to answer this is to consider whether there are any 'cardinal virtues' (Papouli 2018: 5–6) that unite these varying practices into a single overall identity, seen as bound by core ethics rather than discrete skills or knowledge.

Not all the social workers who have engaged with or advocated virtue ethics have sought to produce statements about the core virtues of the profession. Some, such as McBeath and Webb (2002), Houston (2003) or Adams (2009), are concerned to argue for the possibility of a virtue ethics in social work. Others, such as Pullen-Sansfaçon (2010), are focused on using a virtue ethics approach from which to assist new entrants to the profession to grasp their own moral identity and responsibilities – that is, to develop as ethically good social workers. This leaves just two who propose particular virtues as important for social work. First, drawing on Banks and Gallagher (2009), Banks sets out seven core virtues (that she acknowledges as a 'partial selection', 2012: 74): professional wisdom, courage, respectfulness, care, trustworthiness, justice and professional integrity. Second, Papouli offers two lists, the first of which is extensive and includes courage, temperance, self-discipline, moderation, modesty, humility, generosity, friendliness, truthfulness, honesty and justice (Papouli 2018: 5).

Then Papouli (2018: 6) provides a more pared-back list of the four 'cardinal' virtues named in classical Greek and medieval Christian thought: courage, justice, practical wisdom and temperance. These, she argues, provide a starting point from which to consider particular virtues for social work. The reason for this shorter list being a better place to commence can be found in its durability across time and context. Not only are these at the core of the classical Greek and medieval Christian views of the virtues – seen as the virtues that sustain the other virtues – but there is also recent empirical evidence that these moral qualities are widely regarded as core virtues across many cultures (Peterson and Seligman 2004; also see Adams 2009: 87). Certainly, each of courage, justice, practical wisdom and temperance might be considered as relevant to all modes of social work, irrespective of whether service users are individuals, families, groups, communities or even the whole society. Whether other moral virtues, such as respectfulness or honesty, might be considered as relevant to all modes of practice globally across cultures might also be debated.

When we look at the more extended lists of virtues, there are some crucial differences, even in terms of the 'cardinal' virtues. For example, Banks (2012) does not include temperance, while Papouli's (2018) extended list does not state practical wisdom. One reason for this is that Banks (2012) is drawing her statement of virtues from empirical knowledge of social work (as well as the contrast with nursing in Banks and Gallagher 2009). Therefore, her main point of reference is the UK, and the term temperance has a particular meaning in parts of the English-speaking world (associated with the avoidance of alcohol consumption), often perhaps being seen as belonging in a bygone era. In contrast, Papouli's discussion much more explicitly places greater emphasis on the Aristotelian tradition, such that she separates practical wisdom as an 'intellectual virtue' from the others as 'moral virtues' (2018: 5). In their more extended lists, only two virtues appear in both: courage and justice. Thus, even these two advocates for virtues as a central part of social work's ethics present somewhat different portrayals because they ground their understanding in slightly different accounts of virtue. This comparison helps us to note that a search for a definitive list of ethical virtue for social work is likely always to be a provisional process at best, as well as being subject to the influences of whoever is undertaking the task. For this reason, it is not surprising that when it comes to the extended lists of potential virtues, complete agreement is not possible, even when there are ethical virtues on which agreement could be reached.

Yet, does this lack of consistency between key virtue ethics positions mean that there is little point in seeking to identify and discuss ethical virtues that are helpful for social workers to address? Although such agreement may not be possible, the importance of ethical virtues to assist social workers to develop themselves as good practitioners means that the effort is worth it, even if we accept that the process is unlikely to achieve a single final position. Furthermore, some of the virtues that are named by the different discussions can also be seen as closely related, even blending into each other. One example is that of trustworthiness (in Banks 2012: 75), which includes being able to be relied on, not letting others down. Supporting this, the trustworthy person is likely to be someone who is honest and truthful (in Papouli 2018: 5). Indeed, it may almost seem tautological, as someone who is dishonest and untruthful cannot be trusted, at least when this becomes known to others. However, the point of the virtues is that these are characteristics or qualities that are cultivated because they are those of a good person and not because they follow set principles. For the virtuous practitioner what matters is not found in actions as such, but the extent to which valued characteristics have been developed.

Of the various discussions of virtue ethics in social work, only Banks and Papouli provide explanations of the qualities to which they refer. Banks (2012: 74–75) provides short statements about why each of the seven virtues that she suggests are important, while Papouli (2018: 6–7) outlines the 'cardinal' virtues. Combining these lists produces eight virtues, which can be summarised as follows:

- wisdom (professional/practical) – bringing judgement to bear through reasoning about specific situations, in order to make sound choices about possible ways to undertake good professional practice;

- courage – the quality of being able to face challenging, unwelcome and possibly dangerous situations appropriately, through finding the right balance between standing up to everything (which is over-confidence or foolhardiness) and backing down from everything (which is cowardice or fear), of knowing when to run and when to fight, and being able to discern what is worth fighting for;
- respectfulness – being able to see and respond to the inherent dignity and worth of others, grasping the way in which another sees the world without diminishing it simply because it is different to one's own perspective (which is akin to empathy);
- care – the characteristic of relating to other people in such a way as to be able to respond to them with the right amount of attention and with an appropriate degree of interpersonal affect (which has some similarities with the quality of 'warmth');
- trustworthiness – being someone who can be relied on and who can take responsibility for their own judgements, choices and actions, who does what they say (which is close to the virtue of keeping one's promises);
- justice – as a virtue, justice is the quality of fairness, of being someone who seeks to find the right balance between the claims that people may have on the world, including when they have been unfairly treated by others in some way (including by social institutions);
- integrity – the characteristic of being able to hold together other virtues in a way that supports the values of the profession, it can also be seen as the general quality of 'being a good person', someone who is inclined to seek the virtues, a person with a 'moral compass';
- temperance – being well balanced, achieving moderation in all aspects of life (such as knowing when to be serious and when to have fun), of being able to look after the self at the same time as looking after others (being able to seek a work–life balance, to avoid 'burnout' and so on).

REFLECTIVE QUESTIONS/EXERCISES

How are 'social work practices' defined?

What are your thoughts about the statement that there may be many 'social works', as opposed to a single profession of social work?

List the social work virtues posed by various authors, as detailed in the section above.

What are the strengths and limitations of the lack of agreement around a definitive 'list' of virtues for social work?

The other question raised by our comparison with discussions of business ethics, above, is whether professional ethics constitutes a special category of ethics, or if it is part of what might be termed 'everyday ethics'. In other words, do the moral values of professional practice differ from the values that might be held across the whole of a community? On this question Banks (2012: 173–177) argues that it is not a simple binary problem. Yes, to some extent the role of a social worker is not that of a neighbour or family member (for example, it may carry certain legal and policy obligations that are not held just by being a member of the wider community). At the same time, the ordinary expectations that people might have of each other in the wider community (such as asking certain questions, or of being able to make certain decisions about another person) are suspended in a professional encounter only to the extent that the professional role itself requires. For example, a question that might reasonably be asked by a social worker, but not by someone as a neighbour, needs to be in the context of the need to ask that question, and the response that follows can reasonably be expected to be received (heard, recorded, used in decision-making or shared with other professionals) in a manner that is also consistent with the professional's role. Information gained in such a context might be expected not to be shared around in the community without the agreement of the person who provided it, or unless subject to a legal requirement. In this way the respectful social worker is the sort of person who does not breach the values of privacy and confidentiality because of their personal quality of respect for others (not because there is a rule to this effect in a code of conduct). In this the qualities required of the social worker are different to those of the neighbour, for example, seen in whether concern for privacy and confidentiality is greater in the professional role.

Papouli's (2018) analysis agrees with Banks' conclusion, as she draws on the common Aristotelian virtues and then proceeds to show how they might apply in social work. Similarly, Pullen-Sansfaçon (2010) draws on MacIntyre (1985) to emphasise that while the virtues that are relevant to social work are the same qualities that we might find across a society, what sets these moral qualities apart for a profession is the *practices* that constitute the profession, the way in which these practices entail a particular expression of virtues and which distinguishes it from the everyday appearance of the same virtues. Likewise, Hugman (2005a), Bowles *et al.* (2006), Adams (2009) and Reamer (2018) all discuss ethical virtues in social work as expressions of virtues that are widely acknowledged across society. The only sense in which there is a claim to 'specialty' in such values is in the practices of the profession. Here we see an echo of Pellegrino and Thomasma's (1993) apparent assertion that professions are different, because (as we saw above) the difference lies in specialised practices, rather than in any sense of members of any profession being a 'special type of person'. The quality of honesty, for example, may appear to create moral demands on a professional practitioner that are not those of everyday life because of the importance of full disclosure of facts or opinions that in ordinary relationships might be regarded as 'bluntness' or a 'lack of tact'.

REFLECTIVE QUESTIONS/EXERCISES

Do you think social workers have extra ethical responsibilities, in addition to the 'everyday' ethical behaviour expected of the ordinary (non-professional) person? Why/why not?

Learning from practitioners

From the preceding argument, we may conclude that to look for ethical virtues in social work the task is not to engage in a search for special qualities, but instead to investigate the way in which those who are engaged in the practices of social work understand the questions of moral value that they face in their work. For this reason, as outlined in Chapter 1, the discussion in the following chapters and the choice of ethical virtues around which it is structured are guided by research undertaken with social work practitioners as well as the professional knowledge and experience of (four of) the authors.

From the research, both the interviews with 65 social work practitioners around Australia and ten biographies of 'exemplars' in the profession (Pawar *et al.* 2017), we have identified a range of ethical virtues that came out of an inductive reading of the descriptions that these social workers provided of their practices and their understanding of the moral demands that they encounter in their work. Our definition of core ethical virtues for social workers is subtly different again to those offered by Banks (2012) and Papouli (2018), as well as being somewhat longer than Banks' list. It is possible that this is affected by context, including the national origin of these ideas, although other studies that have stated possible lists (Banks and Gallagher 2009; Banks 2012; Papouli 2018) like our own are also greatly influenced by knowledge of international and cross-cultural views of social work practice and of professional ethics and values. As we have already noted, such differences between statements of particular values may be inevitable, like the difference of perspective when looking at the same scene from different angles. In that sense, this range of ethical virtues in social work is offered not to rebut other scholarship, but to add to the understanding that is available to the profession of this important aspect of professional ethics.

The range of virtues that is considered in Chapters 3 to 9 consists of 15 moral qualities. In the order in which they are presented, these are courage; compassion and care; hope, resilience and perseverance (fortitude); justice; humility, respectfulness and dignity; wisdom and temperance; and integrity, including honesty and trust(worthiness). This order reflects stages of our analysis; it does not imply any scaling of importance or significance between the different virtues – all are virtues of social workers that we have identified. The grouping of particular virtues together is influenced by both the practice evidence on which we rely and the existing ideas of virtue ethics in moral philosophy. Although it can be helpful to distinguish different virtues, in order to consider the particular aspects of them, this ought to be seen as a means of analysis and not as a statement about a fixed meaning of any particular virtue. In particular, we can note that the variety of perspectives among social workers and ethics scholars suggests that ethical virtues must be understood in relation to each other.

For example, honesty and trust are very close but they are not exactly the same. A dishonest person is unlikely to be seen as trustworthy, but the person who is honest while lacking respect and dignity might also be regarded as untrustworthy, because there could be no reliable expectation that their words or actions would not simply lead to honesty being expressed harmfully. Slote (1992: 93) has observed that some moral qualities, such as promise-keeping and truthfulness, clearly stand outside the Aristotelian notion of a mid-point between two vices, but it is difficult to envisage an excess of either. They can be deficient – for example, not keeping promises or being untruthful point to a lack. However, what may make them appear to be 'excessive' is more to do with the way that they are used, such as through forgetfulness or a lack of tact.

On this relationship between ethical virtues, Banks and Gallagher (2009: 57) point out that moral dispositions or qualities are not discrete from each other, but can be better understood as clusters or as forming a range. Some of the virtues discussed in this book are the same as those agreed between other social work scholars, especially Banks (2012) and Papouli (2018) – courage, justice, care, integrity, wisdom, respectfulness, trustworthiness and temperance. Others are not (compassion, hope, resilience, perseverance, humility, dignity and honesty), but are taken from our own observations. So, the range of virtues that is considered in this analysis are those that we have reason to think from investigation or experience provide a rich portrayal of social workers and not a definitive catalogue of required moral characteristics. This is not a 'check list', but a discussion of what is seen in the practice contexts from which these ideas are taken. In that sense, the ethical virtues of social workers are situated and must be understood as such.

At the same time, we cannot say that in so far as virtues are situational this should not be understood in a purely relativistic fashion. They are not simply dependent on personal preference or whim, nor can they be defined only in terms of a particular outcome that we might be trying to achieve. The community development practice of Alinsky (1971) appears to do this, in that he advocates using all means to achieve ends that he regards as justified from a particular reading of consequentialism. This view briefly influenced some social work ethics debate, for example, in the form of 'situational ethics' (Goldberg and Elliott 1980), in which it is argued such an ethics might reasonably assert that the good social worker could use dishonesty to pursue a valued outcome, where there is a challenge or risk of being blocked or even disciplined for doing so. Their grounds for this are that any issue where the content or the consequences set the interests of people in different social groupings against each other is political and that in any political context social work ought to take sides with those who are marginalised, oppressed or discriminated against. Such a position is clearly supported by the view of social justice that pervades social work's formal ethical statements (IFSW/IASSW 2018) and many social workers would agree that it is a core ethical and political position (compare with O'Brien 2011).

However, Alinsky's and Goldberg and Elliott's arguments amount to a form of consequentialism that contains a serious problem. It is not that it orders ethical priorities in the relative weight given to concern for different people, which we might agree that everyone has to do in ordinary life (Nagel 1979), but, rather, it does so in a way in which the humanity of some people is disregarded because of their identities and actions. This undermines the moral claims in politically positioned practice, in that it asserts the capacity to pass the very type of moral judgement that enables oppression and abuse to occur. It also emphasises a separation of identity, creating an 'us and them' binary construction in ways that may achieve short-term gains but ignore questions of longer-term community relationships. Therefore, as an ethical argument it is self-defeating. For much of human history, ethics and politics are not separated in this way, so that for Aristotle, Confucius, Locke or Mill, the connection between the good society and the good person is seen in right action (Alexandra and Miller 2009: 268). Principles and virtues are inherently connected and not open to a 'whatever it takes' relativism that so easily slides into 'ethical egoism' (Hinman 2013).

Yet, whatever the political positioning of social work or its adherence to the principle of social justice, the dissonance of deliberately using dishonesty to pursue good ends did not see it become an influential position (Hardina 2004: 598). This is somewhat different to a prioritising of values in which 'being polite' or 'behaving properly' is seen as a matter of etiquette (or aesthetics) rather than ethics. Even when it becomes necessary, confrontation as a tactic, such as in the form of strikes, sit-ins, demonstrations, boycotts and so on, does not have to dehumanise those who oppose progressive change (Pyles 2019: 170). Hardina's (2004) description of the way practitioners deal with unjust situations demonstrates the balancing of responsibilities to vulnerable members of communities with the acknowledgement of the humanity of everyone involved. This can be achieved in ways that reveal the virtues of honesty, care, courage, respect, resilience and practical wisdom. For example, the courage of social workers effecting social change with marginalised and oppressed people can be expressed in various ways that also maintain the virtues of truthfulness and respect towards the humanity of those against whose opposition change is sought. However, those social workers often continue to be seen as troublesome, to be disliked or even to be attacked (compare with practices of Australian social workers discussed in Pawar *et al.* 2017).

In summary, to draw an understanding of ethical virtues in social work practice from practitioners requires that we do not simply record what individuals say about their view of themselves and their own work, but compare across the profession. We need to consider the values that are expressed in relation to wider debates about ethical values – including what can be learned from the traditions of moral philosophy – and at the same time take account of the social, cultural and political context. With that in mind, the following chapters draw on a variety of sources for the ideas that are examined. For example, the questions asked include not only *what are the virtues that social workers refer to in order to explain their practices?*, but *what are the conditions in which these virtues are articulated?*, *what is the range of competing ideas about the ethical virtues of social work practice?* and *how do social workers resolve tensions or conflicts between the virtues that they regard as key to good practice?* It is on this basis that we proceed in the following chapters to examine particular virtues in social work practice.

FINAL REFLECTIVE QUESTIONS/EXERCISES

List three key learnings from this chapter.

Which do you feel are the most important virtues in social work practice, and why?

Do you think your chosen virtues are more or less important across different practice settings? Why/why not?

3

COURAGE

CHAPTER OBJECTIVES

- To define the virtue of courage in relation to social work practices
- To explore the relationship of courage with key concepts and experiences such as struggle, authority, whistleblowing and violence
- To explore expressions of courage across different practice contexts
- To consider how courage may be cultivated in the social work profession

Courage as a virtue

In this chapter we examine the virtue of courage. In social work, courage has various meanings, including boldness, determination, bravery and fearlessness, expressed in the capacity to challenge opposition when pursuing good objectives or to keep going in the face of resistance. It can also be regarded as an innate personal quality, as a conscious commitment to core ethical principles such as human rights or social justice, or it can be a combination of the two at the point where they meet. Two related issues follow from this. First, a virtue such as courage becomes evident in the context of practice. It cannot be observed or understood separately. So, in order to examine it, courage has to be located in practices. Second, as with many of the other virtues that are discussed in the following chapters, ethical virtues are not binary in form—that is, they are not either present or absent in a person. In other words, as with other virtues, courage can be seen to be a quality someone brings to their practice in a more subtle, nuanced way.

Courage may often be considered a virtue that has significance in situations that are especially difficult or demanding. However, from the classic texts in ancient

Greece, India and China, courage has been considered a central quality also in ordinary, everyday life (MacIntyre 1985; Lai 2006). Philosophers such as Aristotle or Lao Tzu discussed courage in terms of boldness and bravery in military combat, and at the same time they also regarded highly qualities of determination and fearlessness in pursuing that which is right in society more generally. More recently, writing as a Christian theologian, Lewis (1998/1942: 115) goes further, to state that 'courage is not simply *one* of the virtues, but the form of every virtue at the testing point, which means, at the point of highest reality' (emphasis in original). That is, it is only when other virtues meet an obstacle that courage clearly becomes evident. At the same time, without courage the other virtues can become conditional on not being challenged. More than this, courage is not the absence of fear; rather, it is the overcoming of fear, including when encountered in its milder forms as uncertainty, hesitation or a sense of misgiving. Courage may also include the development of the capacity to make judgements about what should be feared. The constantly fearful person is unlikely to be seen as courageous, even if she or he manages to overcome this fear on occasions. As MacIntyre (1985: 150, 167–168) observes, by way of the influence of successive generations and ideas from other traditions, such as medieval European monasticism, Aristotle's assumption of the great, high-status man exercising courage has become modified to refer to the ordinary person holding onto her or his commitment to the sort of person they wish to be in the face of difficulties, challenges and threats. For Aristotle, courage had to be tempered by justice to prevent it becoming tyranny. In Chapter 2 we noted that Banks (2012: 73–75) and Papouli (2018: 6) recognise courage as one of the core ethical virtues of social work, with both of their discussions following the argument that courage is central to ethical virtue in practice. In this sense, twenty-first-century social work follows the later European tradition in which achieving justice now requires courage. So, here we consider all aspects and draw out the way in which courage relates to other virtues that are discussed in the following chapters.

REFLECTIVE QUESTIONS/EXERCISES

What are your thoughts about courage as a quality of a social worker?
How important is courage to undertaking social work?

Courage as struggle

We begin by looking at the way in which courage in social work is seen in practice that struggles against opposition or threat in some way. Let us consider a specific (hypothetical) practice situation that models this.

> Jane is a social worker in a non-government agency. She is seeking to support a family who are facing deportation because their claim for asylum under the

> Refugee Convention (United Nations 1951) has been denied by the Home Affairs Department of the national government. The adult members of the family have described the persecution they experienced before leaving their country of origin, in which they were members of a cultural and religious minority. They have also told Jane about the physical threats that they received from members of the cultural majority among their neighbours. The view of the relevant government department is that the family have not demonstrated that they have any reason to think that they will be subject to persecution or threat in the future. Jane is committed to assisting this family to challenge the deportation order, although there are other families also needing a response from her agency, so she has limited time. Politicians and some parts of the media are beginning publicly to criticise the agency and Jane personally, so as the organisation depends partly on government funding, the agency director (with whom Jane otherwise has a respectful relationship) suggests to Jane that she has done what she can and that she should now accept that her time would be better spent on helping others. Jane, however, is convinced that an injustice is being done to this family and chooses to continue to work with them and to challenge the government department.

In this situation, we can see that for Jane to continue to work with the family will require that she has the personal qualities of being able to work against opposition from several sources, including the power of the Home Affairs Department, being attacked through the media and possibly losing the support of her manager, who normally respects her and her work. In such a situation courage certainly involves boldness and determination, as Jane will have to argue on behalf of the family, alongside them in meetings (such as a review tribunal), in a process that will take many hours and will involve responses that seek to block her efforts.

There are also aspects of bravery and fearlessness involved in this situation, as Jane may well experience personal criticism and opposition, not only from public sources (political and the media), but also from colleagues who she might normally expect to share her values. Depending on the way the agency functions, she might even find that she is instructed by the director to reduce or refocus her involvement. So, by continuing to work intensively with the family she would be failing to implement a decision that the director has the formal authority to make. In a neo-liberal policy environment, moral courage may be required in order to maintain the commitment to pursue justice with and for service users (McAuliffe *et al.* 2010; Hyslop 2018; Bay 2019; Papouli 2019b).

All of these factors together create the particular circumstances in which Jane expresses courage. If she takes the Aristotelian position that courage is the mid-point between foolhardiness and cowardice, she might pause to consider the point at which she is going to conclude that it is time to stop. However, if she sees courage as the 'testing point' of her compassion, resilience, perseverance, justice and integrity, she may maintain her engagement with the family to the end of the review process and, if unsuccessful, until deportation. The complexities of how this proceeds include Jane's courage (for example, in her capacity to deal emotionally and psychologically with criticism and verbal attack), as well as the way in which others act, including immediate colleagues who may not be seen as 'oppositional' in other situations but whose support may be lacking in this instance.

One of the ways in which courage is enmeshed with practical wisdom and integrity lies in being able to make sense of these surrounding factors while 'remaining one's self' in the situation. As we noted in Chapter 2, practical wisdom in such circumstances enables the practitioner to know which issues are worth pursuing or defending, when to take a stand and when not to do so. For example, we can conclude from this situation that Jane's character inclines her to engage with issues and situations in which she thinks people are treated with a lack of compassion and justice, in which the lack of these qualities will likely lead to a harmful outcome for the family members. Here, practical wisdom enables Jane to express her own compassion and justice, by giving her a reason to be courageous.

At the same time, courage is not only about being able to see what is the right thing, but also the moral strength to pursue it when to do so is risky. It involves someone being able to face reasonable concerns about their own safety, whether physical, psychological, emotional or social, in order to maintain their integrity. Courage is not about just doing something that is demanding; rather, it only finds its full expression when facing fear. In these circumstances, Jane can reasonably feel concern that by pressing on with active support for the asylum seeking family she will be subject to potential threats, for her reputation and even, ultimately, a job that she values very highly. Yet, as someone with courage who also has integrity, Jane does not want to be the sort of social worker who backs away from pursuing the right objectives. For her, integrity can be understood as the quality of bringing ethical virtues together, so that the moral self remains whole.

In this situation the hypothetical social worker, Jane, is engaged in the practice of advocacy, including awareness raising, helping marginalised or disadvantaged people to represent their voice to those with greater power or, where necessary, acting to represent a voice for them, and promoting better programmes and policies. Advocacy is also part of community work and policy development, in which the same question of opposition must be addressed (Kenny and Connors 2017: 438). In all these forms of practice there is a strong likelihood that conflict of some type will ensue and it should be recognised that even if the notion of threat is not about someone's own immediate safety, working with conflict in itself requires comparable forms of courage. Many people do not like conflict and have consciously to seek to overcome a natural inclination not to provoke it. For example, nine of the ten in-depth biographies of Australian social workers discussed by Pawar *et al.* (2017) demonstrate courage related to struggles of various kinds, including advocacy, community work, policy practice and even as managers seeking change within organisations against hostile political attitudes. All describe some sort of tension with others – even overt conflict – including with social work colleagues or those in other professions. In that phase of the research undertaken by the current authors, examples of courage against these types of opposition included pushing for changes in adoption law and judicial decision-making, and getting colleagues from other professions to change their ways of working when this could have been regarded as a criticism of those colleagues. In these circumstances, courage to accept that conflict is a necessary part of practice is a quality that enables social workers to overcome continually living and working with debates, disagreements, disputes or even overt aggression. There is also courage in deciding to leave a job when it appears that a struggle against injustice means that the conflict is with your own professional colleagues and where staying would involve an unacceptable compromise of a person's own values.

REFLECTIVE QUESTIONS/EXERCISES

What challenges is Jane facing in the above case scenario?
If you were a social worker in Jane's place, what would you do?
Reflect on what sort of courage is needed in the case example, and what can hinder courage.

Courage as exercising authority

Often social workers are required to exercise authority in work with individual people and families, and in the supervision of colleagues. Courage in practice with individual people and families may at first appear somewhat counter-intuitive. Social workers exercise power and authority in relation to service users, sometimes to quite a considerable degree (Hugman 1991; Ferguson 2001; Ferguson and Lavalette 2004; Smith 2008; Gilbert and Powell 2010; Belcher and Tice 2013). Although some analyses regard power as inherently structural, others allow that it is also a more fluid facet of human relationships. The virtue of courage is more easily seen in relation to structurally focused practice, where courage is framed as struggle against oppression. What is harder to reconcile is when courage is called on in exercising authority and power over others – indeed it may even be that the dynamic of such authority relations is what makes it necessary to think in terms of courage, as much when exercised in relation to service users as to colleagues. In such relationships social workers need to have courage to exercise power and authority well, when they are seeking the good of the service user but often against that person's objection and criticism from others. Exercising power and authority is often contested.

Examples of the exercise of authority and power can be found in child and family social work, mental health, criminal justice and in the direct provision of material services (social security, housing and so on). While the specific service areas vary between countries (for example, social work services are found in the government social security department in Australia but not in the UK), the implications for the ethical virtues in practice are comparable. The following hypothetical practice situation provides a basis to consider this dimension of courage.

> Gloria works in a busy urban social work office, providing government-mandated services for children and their families. Concerns have been expressed by a school principal about a family with whom she is working, because teachers have observed bruises on a six-year-old girl, Faye, that are not consistent with her explanations that she fell in the school playground. Gloria is from the same migrant community as the family and has developed a good relationship with the girl's parents, acting as an ally in dealing with parts of the social welfare system. However, Gloria also has worked with the parents to encourage them to find ways of disciplining their children that do not cause concerns about the use of physical force. Although Gloria appreciates the stresses of being migrants, she also does not accept that physical chastisement is the only cultural norm on which the family can rely. She wants to ensure that they understand the local child welfare laws. When she arrives at the school and is able to talk with Faye, because of what Faye tells her about how she really got the bruises from her father hitting her, and how he has been behaving to all the children in the family, Gloria is concerned that the family situation is becoming more conflictual. Her next step is to talk with the parents, but she is aware that she has a responsibility to ensure that the children's rights not to be physically harmed may be in conflict with the parents' perceptions of their needs and rights. She decides that although it will affect their working relationship, she will have to challenge the parents directly about the use of physical punishment. Gloria will also have to make it clear the law means that she now has a responsibility to make sure that it is safe for the children to remain at home with their parents, and that in doing so she is acting on a statutory obligation for her department to receive a report of concern (in this case from the school). Appreciating how hard they have worked to settle in a new country, and the problems they have faced in the past with employment and housing, Gloria knows she will find this particularly difficult, but that she has to find the courage to have this conversation, as she thinks the children are possibly at risk of harm in the future if it is not addressed and that avoiding the issue will not be helpful in the longer term.

What Gloria is facing is the challenge of confronting acts that are detrimental to the safety of Faye and her siblings, but where she has developed psychological and emotional bonds with Faye's parents and now has to say things to them that they may not find easy to hear. There is courage in doing what is needed to make Faye's needs and interests to be physically safe paramount, and in the process risk the relationship that has been carefully built up with Faye's parents. This is where 'controlled emotional involvement' becomes a mediating capacity in social work practice (see Chapter 2).

Social workers have debated the rights and wrongs of exercising authority in this way for a long time (Howe 2014). Usually this debate centres on the values of human rights, with a contradiction seen between the rights of parents to bring up their children in the ways that they value (including according to cultural expectations) and the rights of children to grow up safely, without being harmed. Although physical harm is at the centre of Gloria's concerns at this time, she is also conscious of the emotional, psychological, cultural and spiritual harm that may result from the actions of the parents, but also from her own actions if, for example, a substantiated situation of abuse leads to Faye and her siblings being taken into out-of-home care. For this reason the courage that Gloria brings to this encounter will be better balanced with the virtues of compassion, care, humility, justice and wisdom. However, the courage to respond to the evidence of physical harm is at the centre, because without it there is the possibility that compassion, care or a sense of justice can become distorted and a social worker might fail to recognise the interests of each person (Phillips and Hugman 1999).

The circumstances in this child and family welfare situation share with social work in criminal justice the factor of engaging with people who have been alleged to have contravened the law. In the criminal justice field, whether in community corrections or probation (as it is variously called), or in prison welfare, social workers are often required to exercise some degree of courage. This can be in facing a person with the implications of their actions, or it can be in recognising and responding to the way in which other factors affect who is and who is not sentenced within criminal justice systems. In other words, the virtue of justice can be challenged by the way that the principle of justice plays out in the court and prison system. Those who are convicted of crimes disproportionately experience poverty, homelessness and mental health issues. 'Race' and cultural issues are also a factor in many countries, as in the proportions of ethnic minority people who are convicted compared to members of the ethnic majority. For example, based on existing data, Indigenous Australians are the most incarcerated population group in the world, relative to their proportion in the overall population (Anthony 2019). Courageous social work practice in such situations may take the form of seeking the recognition of the humanity of those who have been convicted, for example, in a social worker ensuring that the person in prison has a voice in situations where colleagues from other professions do not think to include that person, as well as joining with community groups and others to argue for changes in sentencing policy.

REFLECTIVE QUESTIONS/EXERCISES

What challenges is Gloria facing in the above case?

If you were a social worker in Gloria's place, what would you do?

Reflect on what sort of courage is needed in this example, and what can hinder that courage.

In the mental health field courage can be seen in a different form. The exercise of authority by social workers in some countries takes the form of involvement in assessment, which may contribute to a person who has been diagnosed with a mental health problem that causes them to be a risk to their own or someone else's safety to be compelled to enter a psychiatric hospital and receive treatment, or be required to receive treatment against their will in the community. Under such systems it is usual for social workers, as well as other comparably authorised health professionals, to take responsibility independently of the medical diagnosis. That is, the social worker may assess that the person who has been diagnosed is not causing a risk and decide not to support their admission to hospital or compulsion to receive community-based treatment. However, evaluating risk in this way is not a precise calculation and is achieved through the use of practical wisdom (Webb 2006; Hardy 2015). For this reason, a social worker may need to be courageous in their decision-making because, as in the child and family welfare or criminal justice fields, risk is not easily determined. As Hardy puts it (2015: 181), risk is a social construction but this does not mean that it is any less real in the effects that it has on the lives of all those who are involved, nor in the public's perception when something later 'goes wrong'.

As in child and family welfare and in criminal justice, social workers in mental health develop particular types of relationships with those who receive their services. In the specialised sub-field of forensic psychiatric social work, mental health practice can concern the exercise of authority in a similar way, such as being part of systems of compulsory hospital admission or community-based treatment. At least in some parts of the world, social workers are able to seek to work in partnership with those who experience mental health problems in supportive and advisory roles, or as 'allies' (Postle and Beresford 2007; Bland *et al.* 2015). There is still a form of power and authority here, but it is that of expertise rather than coercion (Hugman 1991).

However, maintaining roles in which power and authority are exercised carefully and for the long-term benefit of service users is complicated and demanding. In particular, this arises from the contradiction between exercising 'power with', 'power for' and 'power over' and how these distinctions can be achieved (Hugman 2017: 72). Working with the complexities that this generates requires courage, as social workers are constantly addressing uncertainty and the very real risks that go with this.

Courage in 'ordinary' situations: everyday practice

So far we have considered the way in which courage appears as an ethical virtue in practice that struggles against injustice or opposition and practice that exercises authority. There are also other aspects of practice that are less obviously contesting these sorts of challenges, but where nevertheless social workers see their work as requiring them to have courage. Banks and Gallagher (2009: 175) quote a social worker commenting that visiting people in their own homes is a challenge, which he summarised in the question, 'how hard should I knock?' This is a comment to which many social workers will relate. In another context, it might be challenging for a social worker to question a service user's way of expressing her or his own sense of right and wrong, which can be seen in this anonymised description of one encounter provided by a social worker in a research interview.

> So I spoke to this young man about being rebellious. I'm not often confrontational with clients, but I challenged him and said, 'are you really rebellious and just want to do things your way?' And we spoke a bit about that. He got it that [he thought] if there was anything wrong then he was going to fix it. So he would take on other people's fights and get thrown out of boarding houses [...] whether they were right or wrong was irrelevant, he liked the fight. The social justice fight. But [...] it wasn't a good thing to do; it got him into trouble. So we spoke about that and I asked him if he had any other way of rebelling [...] that would perhaps be less destructive. But [challenging someone] is a little bit scary and I don't know if I would plan to do it, just do it when I thought it was right.

At the end of this statement the social worker feels uncertain, even that it is risky to challenge the service user about his actions. It is not a confrontation of the man's rebelliousness against perceived injustices, but rather a concern about the impact of the way this was being done. In this way, the courage of the social worker is enabling compassion and care to be expressed while at the same time seeking to be respectful towards someone's own values and qualities. As discussed previously, practical wisdom plays a part in enabling courage to be expressed, so that the social worker relies on a sense of knowing when to speak and when not to do so; but there is courage in saying what needs to be said when the moment is right. Knowledge of the self as a practitioner plays an important part in this, as well as good judgement about the young man.

In another statement of the way in which courage is demonstrated in everyday practice, when asked to explain what she meant by courage a different social worker observed:

> Last week I was with a client of mine where [...] I've seen her for a year or so and there's a strong alliance, but I felt I needed to call her on some behaviour that she needed to take responsibility for [...] we had a conversation about that. I could've let it go and kind of gone along but I don't think it would've been helpful to her in the long run at all.

Again, the practitioner is giving a sense that courage is found in not taking the easy way, but rather of doing what is right, even if it is inconvenient or uncomfortable.

A third example of courage in practice is in seeking to assist people in situations where there actually appears to be a low likelihood that their lives will improve. The following reflection from a mental health social worker conveys this.

> Most challenging is accepting that there are some people I work with who aren't going to have better lives than they have got now, that there is actually nothing I can do. So being able to [...] it's not about lowering my expectations [...]. There's no nothing. This is how they're going to be. To recognise that just sitting with them in a room for an hour is wonderful for them, even though it feels like I'm not actually doing anything. I am doing something by being there. That's been a big challenge because I want to fix the world and you just can't.

The social worker in this vignette describes this as an example of the most challenging type of practice that they do. Courage in this situation is seen in the boldness to recognise that the opportunities for 'fixing the world' are not always there and that 'sitting with someone' is sufficient. It is courageous because this practitioner has to overcome their own resistance to 'not fixing things' as providing what the service user wants and needs. As ethical responses, moral distress (Lynch and Forde 2010) and moral outrage (McAuliffe *et al.* 2010) to oppression, discrimination and disadvantage are not only experienced in relation to extreme situations, but may also be encountered in the intractability of the issues faced in some people's lives where all that is available is being provided.

Davies (1994) argues that some social work is about what he calls 'maintenance'. By this he means that it may take active intervention to assist someone to lead as good as life as possible, but that this can be about helping someone to keep going rather than about achieving large changes. It is not about 'leaving things as they are', nor is it about neglect, benign or otherwise. Rather it is a recognition that in some circumstances the effort of keeping things as they are, so that they do not deteriorate, is what is required. Not all social work is about maintenance in this sense, but when this is the case the courageous social worker can admit that this is what they have to offer and can sit with the person in their situation.

REFLECTIVE QUESTIONS/EXERCISES

What were the elements of courage in the above three practice examples?

Assume that you are the social worker in each of the three practice examples above and reflect on what you would do.

Courage in unusual situations: whistleblowing

The issue of whistleblowing is addressed in several studies of social work ethics (Banks and Gallagher 2009; Banks 2012; Parrott 2014; Papouli 2018; Reamer 2018). Historically, there have been relatively few instances of social workers 'blowing the whistle' on malpractice in their employing agency. It involves making public a concern that certain actions by professionals or events in an agency are harmful or corrupt. Whistleblowing is somewhat different to the lack of support to challenge external injustices or obstacles that we considered above. In a situation that leads to whistleblowing, the person or people against whom courage becomes expressed are one's own colleagues, even possibly a senior person such as a supervisor or line manager. At the same time, in speaking publicly about the perceived wrongdoing, the whistleblower is usually acting against agency policy and will most likely be seen as acting badly themselves by 'being disloyal'. The whistleblower may even be accused of breaching agency policies or breaking laws by speaking out. Such counter-accusations can lead to whistleblowers losing their jobs before the truth of their concerns is tested; but even when their concerns are substantiated, obtaining work in their profession may become more difficult as 'no one wants to work with a troublemaker' (Parrott 2014: Trappenburg and van Beek 2019).

Let us return to Jane's situation presented at the start of this chapter. Jane's decision to continue working in advocacy with the family seeking asylum displays courage more clearly if we find that she is aware of the following Australian example.

> In 2014 human services professionals, including social workers employed by Save the Children, were providing support services for asylum seekers who, under Australian law, had been placed in a detention centre in Nauru (a very small Pacific island nation that receives income for providing this facility). These professionals had information about the deteriorating mental health of detainees, including that expressed in acts of self-harm, of abuses against children and violence against women that were occurring in the centre. Deeply concerned that other ways of securing safety for these vulnerable people had not succeeded in changes to detention policy and practice taking place, the professionals decided to make statements to the Australian press. As a consequence, they were all removed from Nauru and Save the Children was replaced by another service provider. The professionals were accused of breaches of contract and of manufacturing complaints, even of actively encouraging detainees to self-harm in a bid to be allowed to go to Australia. Support from sections of the public, advocacy groups, professional bodies, politicians and the press led to the Australian Government, apparently reluctantly, creating a public inquiry. This inquiry exonerated the professionals and cleared them of wrongdoing (Moss 2015). However, the Australian Government refused to accept these findings and launched a review of the first inquiry. The review found that the first inquiry had been correct in its process and conclusions and substantiated its findings (Doogan 2016). While no further action was taken against the professionals, the Australian Government has continually refused to accept this outcome or to apologise for the vilification of the professionals who were directly involved and of Save the Children.

The professionals who were involved in this situation, and those in Save the Children in Australia, certainly can be said to have displayed courage in doing what they knew to be right in the face of powerful opposition and potential harm to themselves.

Had Jane considered this series of events, and the fate of the colleagues who 'blew the whistle', before making her decision to continue to work as an advocate with the asylum seeking family, she might have given some thought to its implications for her own employment. It is not that she is acting as a whistleblower, but the agency that she works for is partly funded by government contracts and the implication for the agency's future funding opportunities could reasonably be a factor in decision-making. In other words, the fearfulness of losing a job, or of others losing theirs because of continued challenges to government policy, might constrain the decisions a social worker makes about how to do their work. The courage it takes to be a whistleblower is considerable. The consequences of what often happens to whistleblowers affects professionals more widely and can create a form of fear that must be overcome in less overtly threatening circumstances.

REFLECTIVE QUESTIONS/EXERCISES

Consider the above case example.

If you were a social worker in this situation, would you raise questions about unjust practices? Discuss your reasoning.

What is the role of courage in situations of 'whistleblowing'?

Courage in unusual situations: violence

Most instances of courage being a central virtue in social work are those in which the practitioner faces challenges that are social, psychological, emotional or perhaps spiritual. However, there are a smaller number of occasions in which physical threats or even direct violence may be encountered. At the extreme, some social workers have lost their lives or been severely injured while at work. For example, in recent decades social workers have been killed in several countries, including Aotearoa New Zealand, Australia, the UK and the USA in the course of their work. Those engaged in social or community development have also been harmed. Recent instances include Finardo Cabilo, who was murdered in 2009 while working against human trafficking in Malaysia, and the then Secretary-General of the Kuwait Association of Social Workers, Feisel Al-Sane, who disappeared when Iraq invaded Kuwait in 1990 and was still missing a decade later, in both cases because they publicly stood up for professional values and objectives (IFSW 2002, 2009). Social workers in areas of conflict, such as war or civil unrest, also face physical threats, sometimes on a daily basis (Lavalette and Ioakimidis 2011).

Much more frequently, social workers may be subject to threats, intimidation and even vilification in the press or social media (Lymbery 2001; Cordoba 2017). Whether in situations of actual harm or of threat, courage is seen in three forms. First, is in simply turning up to work. To be able to keep going when either you or a colleague have been harmed or threatened requires courage to overcome the reasonable fear that this can create. Second, courage is seen when social workers not only remain in their roles, but also when they take on work that carries risk. Fear is an emotional and psychological reaction as much to something that has yet to happen as it is to what already has happened or is happening; so, beginning a new task that carries such risk requires overcoming fear. Third, in the longer term, maintaining a commitment to the wider values and goals of social work also requires a form of courage, in that it calls for a commitment to overcome the less obviously challenging way in which constant criticism and apparent lack of public support for social work undermines confidence.

Understanding the impact of violence, aggression and intimidation in terms of courage may bring with it an unintended consequence, namely that when social workers are not willing to keep going in the face of what we have identified as fear, this passes negative moral judgement on their decision. That is not what is intended. Moral philosophy has long understood that there are qualities we might admire but which cannot be demanded of someone. Expressions of these qualities are known as a 'supererogatory', meaning that they 'go above and beyond' what can be expected of anyone in this situation. Beauchamp and Childress (2013: 45–47) define four conditions for regarding this degree of courage as supererogatory:

1. it must be both optional and possible, that is, neither demanded nor excluded by normal standards;
2. it must exceed what is demanded by normal ethical standards;
3. it must be done deliberately and for the benefit of others – that is, it cannot be either coincidental or done for self-interest;
4. it must be good in and of itself and not done only with good intentions.

So, for example, a social worker who turns up to work and then is physically threatened by someone later in the day has not necessarily shown courage simply by being at work that day. However, if they show up to work the following day, even though they could have taken time off, to visit another service user who has an urgent matter to resolve, that might be regarded as supererogatory. Nevertheless, others might see this as overcompensation to the sense of threat and so, from an Aristotelian perspective, a vice (even if understandable). Certainly there is more than a moral hazard in over-valuing courage when this might lead to someone taking unnecessary risks. How such a response is regarded depends on the circumstances of each instance and the criteria Beauchamp and Childress offer provide a basis for examining any risks in order to consider whether they are supererogation or foolhardiness.

REFLECTIVE QUESTIONS/EXERCISES

Do you think it is worth demonstrating such courage in the face of violence? Reflect on your reasons.

Cultivating courage

In general terms, there is a longstanding question as to whether ethical virtues are innate qualities of individuals or if they can be learned or developed in some way. Even if we accept that most of who we are is developed through experience within the context of our lives, is there not also a sense that we are born with some elements of personality already defined? Courage would appear to be a particular case in point, as it can often seem that a person either is or is not courageous, but that this only becomes clear in the moment when their character is tested.

To address these questions it is necessary to remind ourselves about the nature of virtues. MacIntyre (1985) clearly argues against this being a sufficient way to understand ideas in general and notions of virtue in particular. In other words, they are not simply expressions of a specific point of view, a matter of taste or of how a person feels. Rather, a community (which here includes the community of a profession) can develop and sustain shared views about those moral characteristics that are considered virtuous. It is not that everyone shares them, at least not equally, but these are qualities that are generally valued. Nor is it necessary that a virtue is seen to be 'pleasant or useful' (MacIntyre 1985: 160); what is relevant is whether it contributes to the practices that are generally regarded as characteristic of what is good in the profession.

Courage is an ethical virtue that we have so far argued is relevant both to the most challenging situations faced by social workers and to obstacles encountered in everyday practice. In some instances we might say that courage is more easily seen in others than in oneself. The examples of 'ordinary' practice that are discussed above are examples of courage because the social workers who provided them were explicitly asked to reflect on their work in this way, whereas the illustrative descriptions of Jane, Gloria and the account of social workers and other professionals involved in the detention centre on Nauru are all presented clearly as statements of courage because they are viewed from an external standpoint.

From this, we can see the virtues generally, including of courage, as qualities that can be developed but not taught in the way that a scientific theory might be taught. The model of ethics education provided by Pullen-Sansfaçon (2010) for students who are beginning a social work career illustrates this. Using 'Socratic dialogues' in a university programme, Pullen-Sansfaçon's model helps students to develop a process of practical reasoning about challenging situations. Through this method, students learn to ask themselves critical questions about their ethical responses to the demands of practice. Courage is an ethical characteristic that Pullen-Sansfaçon identifies as being developed through this process and notes that students identified their own learning as relevant for their practice (2010: 413). Papouli (2016, 2019a) also proposes active learning of this kind, with the particular method of asking students to engage critically with the practices they are encountering.

Groessl (2017: 73) argues that an important ethical issue in social work is silence about ethics in practice, whether of practitioners themselves not sharing concerns or supervisors not recognising the need to support and encourage colleagues to do so. Without this happening, Groessl suggests (2017: 75) that social workers are more likely to experience burnout, secondary trauma or compassion fatigue. However, she recommends that there are two ways in which social work can seek to avoid this (2017: 77). First, preparing new entrants to the profession by familiarising them with the complexities and uncertainties of practice can assist social workers to gain insight into the moral ambiguities that are inherent in the role, in the manner described by Pullen-Sansfaçon (2010). Second, providing support and ongoing professional development enables courage to be sustained and even to grow over time. Initial education cannot be the complete answer to a working life that is full of challenges, even if it does potentially set the foundations. As Zimbardo argues (2007), people are not uniformly good, and it is empirically shown to be possible for someone who starts out with strong positive moral ideals to become corrupted by context and experience. Conversely, in an environment that positively supports ethical virtues it is possible for someone to become more rather than less courageous as they have this virtue modelled in the practices of others, acknowledged verbally and incorporated into formal agency policies. This might even include the courage to be able to make difficult decisions that for structural reasons are choices between actions that each have negative consequences for someone – that is, the courage to choose what is 'least bad' and work positively with that decision (McAuliffe 2005; Weinberg 2010).

The idea of 'cultivating' a virtue such as courage is a conscious metaphor – it needs actively to be grown, not only to be left alone. It must be nurtured or it can wither. Although this is the case for all the ethical virtues (MacIntyre 1985; Hursthouse 1999), it may seem less intuitive in relation to courage that such a quality can be developed or diminished. However, as we have attempted to show, courage is not binary, capable of being only present or absent. Rather, it can be more or less present or absent. It can be enhanced or diminished. What we are suggesting is that because of this, courage has to be attended to explicitly. Such attention requires that initial professional education and life-long learning should contribute to courage being recognised as a key virtue. At the same time, if courage is to be regarded as part of the ordinary, everyday qualities that enable social workers to practise well, supervisors and managers must enable them to do so. In particular, supervision is a crucial opportunity for social workers to reflect on their ethical responses to the demands of practice, as well as for supervisors to model and encourage ethical reflection.

Banks and Gallagher (2009: 191–192) also recommend that social workers learn more about courage from the world around them, such as through reading examples of courage in literature, or from a discipline such as social history. They suggest that accounts about the lives of 'ordinary' people, as well as those of people who might be regarded as 'heroic', can be of interest in examining courage. In this way we can find 'moral exemplars' of courage in everyday life as well as in 'great' stories. Exemplars of courage come in many forms, including confrontations with authority that is being misused, standing up to bullying and holding on to what is seen as the right view when this threatens a career, alongside quietly maintaining commitment in a situation that appears intractable and in choosing a field of practice that is unpopular and unlikely to lead to career advancement (Pawar *et al.* 2017). Here, too, critical reflection is important as the way by which the lives of other people can illuminate our own lives, as we compare our own responses to those of exemplars.

REFLECTIVE QUESTIONS/EXERCISES

Do you think the quality of courage can be learned and increased?
Reflect on what you can do to further develop the quality of courage.

Conclusions on courage

Courage is seen in social workers engaged in many different forms of practice. When it is absent it can lead to professional silence on matters that ought to be of explicit concern. In such circumstances it can be the source of 'moral distress' or 'moral outrage', which in turn can be a powerful motivator for good practice (McAuliffe *et al.* 2010; Papouli 2019b). Yet, courage can also be necessary to face the dilemmas found in the limitations of practice, for example, where intractability of problems means that a social worker has to work together with a person just to prevent their life from deteriorating.

A central thread of this discussion has been that courage is not a quality that is seen only in exceptional social workers. Rather, it is an ethical virtue that may be expected of all practitioners and, indeed, is so often evident in their everyday work. It is not simply either present or absent, but can appear in differing degrees and different forms. Courage can be cultivated and must be sustained because it can also be diminished when the prevailing practice culture inhibits or suppresses it. Ultimately, leaving a job that is morally toxic is courageous – perhaps even more so if it is possible to make it clear why one is departing to people who can do something about it.

In this chapter we have explored the ethical virtue of courage from the perspective that it is central to all ethical virtues in social work. In the following chapters we examine other qualities that give courage a purpose and help to shape its expression in practice.

FINAL REFLECTIVE QUESTIONS/EXERCISES

What have you learned from this chapter? List three main aspects of learning.
How will you translate your learnings into action/practice? Write an action plan.

4

COMPASSION AND CARE

CHAPTER OBJECTIVES

- To define the virtues of care and compassion, and their relationship with one another
- To explore expressions of care and compassion across different practice contexts
- To consider how compassion and care may be cultivated in the social work profession

Introduction

The ethical quality of compassion can be said to have something of a mixed reputation in social work. Likewise, the meaning of care is not uniformly agreed. In this chapter, therefore, we begin by looking at the debates that have occurred around these ethical virtues and set out arguments regarding critical interpretations of them as crucial for placing them among the vital qualities of social workers.

Compassion and care are closely related, which is why they are addressed together here. At the same time, they are somewhat different in their form and their implications for social work practice. For this reason they are first addressed separately, establishing for each an explanation of how they constitute an ethical virtue, and are explored in more detail in relation to the ways in which they inform social work. The later part of the chapter then provides a comparative examination of the way in which they interact and support each other

Compassion is distinguished from the ethical 'vices' of indifference and pity, and is shown to be a significant factor in the highly valued capacity of empathy.

Compassion may also be considered as the opposite of disgust, condemnation and rejection (Nussbaum 2001). By inclination, the compassionate social worker draws on emotional intelligence to understand how they could be drawn to such negative responses in the face of the apparent objectivity of other people's choices and actions. Instead, compassion leads the social worker to bring empathy to their engagement with that person, in such a way that the humanity of the other is grasped. In this way, emotional as well as cognitive intelligence can support this dimension of the moral character of the practitioner.

In the discussion of care as an ethical virtue, we identify its connections to the 'ethics of care', while also explaining why that approach is not a virtue ethics. Care as an ethical virtue is articulated as a quality of the practitioner and not as a principle or a goal. However, those ways of understanding care can be shown to relate to the way in which care is expressed, in that the virtue of care enables social workers to pursue such values and to avoid their opposites. It also provides insights into the longstanding contention between care and control as the outcomes of social work practices (Day 1981; Howe 2014). In particular, this discussion raises questions of the limits of compassion and care as ethical virtues in social work and shows why and how they must be understood in relation to other virtues, such as courage, justice, humility and respectfulness.

Defining compassion

Nussbaum (2001) describes compassion as the ethical response to the plight of another person that sees in it the problems of the human condition, particularly as this involves oppression or suffering. Where pity implies a sense of superiority on the part of the person who expresses it, compassion comes from a sense that problems and issues arise simply from being human; this is something that could happen to any one of us, if we were in that situation. It is not that we have to have experienced the same misfortune, only that we can see we share a possibility of doing so. This is what makes pity different, as a sense of superiority leads to an attitude that 'such a thing could never happen to me', implying that it is because of qualities I have that the other person lacks. So if I pity someone, I 'feel sorry for' them; I do not 'feel sorrow with' that person.

Nussbaum's discussion introduces two dimensions to compassion. First, she states that in order to be compassionate someone must have the capacity to grasp the seriousness of the plight experienced by the other, which she refers to as the 'size' of their misfortune (2001: 307). It is not that the person who has the problem experiences it as important; compassion is in the ethical response of the one who is seeing it from the outside. Someone can regard their difficulty as trivial, while everyone around them considers it serious. Equally, someone might think of their own situation as a misfortune, while everyone around them considers it trivial. In this understanding, compassion is the ethical quality that inclines someone to make good judgements between unfortunate life experiences.

However, second, Nussbaum presents a problem for compassion in social work in that she observes a long history of compassion being seen in some way as related to a sense of whether the other person deserves their problem (2001: 311–312); in other words, whether they 'are the author of their own misfortune'. At the same time, on this point Nussbaum is quite nuanced, in that she notes that in general since the late twentieth century there has been a marked shift in view since the Aristotelian period. Although she notes that some cultural and some political perspectives might see some plights as the responsibility of the person suffering them – such as in some political attitudes about poverty as avoidable through having the will to work hard – many others not only see unemployment as arising from the lack of job opportunities, but even when it might be ameliorated by personal determination that any loss of will is itself part of the plight of poverty (Nussbaum 2001: 313–314).

This latter view of compassion is attractive for social workers, for three reasons. First, social work addresses human life experiences that can be widely agreed as serious plights: physical and mental illness, including alcohol and other drug misuse, as well as health issues that arise from old age; trauma, including from war and forced migration; personal abuse and victimisation, including gendered violence; 'natural' disasters; poverty, unemployment and homelessness; relationship problems; bereavement; exclusions arising from disability. Social work also is increasingly seeking to be responsive to the ways in which such plights are exacerbated by structural oppression and disadvantage, through issues of gender, sexuality, 'race', ethnicity and culture, disability, old age and so on.

Second, social work as an activity is directed not towards those who have problems, but those who require assistance to address their problems (England 1983: 13). Yet, although only some people need to be helped, finding oneself in that situation potentially could happen to anyone. Even if life choices or personal actions have contributed to the plight, social work values direct attention to assisting people to resolve their difficulties. Help is not denied because of blame. It is not that social work seeks to diminish the dignity of the people concerned by denying them agency, rather to assist them to regain that dignity by being able to exercise their agency. This is understood as the objective of 'empowerment' (for example, Adams 1990; Levenson 2017). For example, we might see this in practice with people convicted by courts of law, where the focus is on helping the person to cultivate their capacities to avoid or overcome their problems in the future, which actually respects their agency. Insofar as everyone learns to cope with the problems of life and to take responsibility through education and practice, this social educative role of social work can enhance rather than diminish the social and moral standing of the person who receives it.

Third, as ethical compassion is the impulse to stand alongside the suffering of others, this virtue connects with and contributes to the achievement of other ethical values of practice. For example, just as the virtue of humility is antithetical to the expression of indifference or moral judgementalism, it is also contrary to pity, because (as we discuss in more depth in Chapter 7) it enables social workers to avoid a sense of moral superiority by reminding them that they share common humanity with those they are assisting. In previous chapters we have noted that what helps to define social workers' practices as 'right' is found in the social context and so in the objectives of their actions, which are value-laden. At their broadest, these objectives can be seen as promoting the dignity and worth of all people, and the achievement of human rights and social justice. Compassion expressed as attention to serious plights and assisting those who need help without seeking to use blame enables social workers to pursue the objectives of their profession. Halifax (2012: 228) argues that the sense of common humanity combines with such professional goals to produce the disposition to pursue the relief of suffering that is central to compassion. From this, we may say that social workers and practitioners in other comparable professions find their motivation in this ethical quality.

Halifax (2012) addresses compassion in the field of palliative care for people who are dying as a result of cancer. She argues that it is particularly relevant because such an illness can occur to anyone and the seriousness of its plight is considerable. Yet cancer may sometimes be seen as a condition over which people bear some blame, for example, when smoking tobacco or consuming alcohol or other drugs has contributed to the cause of the illness. In countries where screening is available, not getting screened can also be seen as culpable. Even when there is no sense of blame, the bodily effects of cancer can evoke reactions of disgust or numbness to the suffering of the other person (Halifax 2012: 231). For these reasons, overcoming the possibility of blame in health-related practice, as well as disgust or numbness, may be as relevant to achieving compassion in palliative care as it is for the treatment of the young person who has harmed her or himself, contracted a disease such as HIV or hepatitis C through unprotected sex or sharing needles, or avoidable physical injury. In these ways, the importance of compassion as an ethical virtue is as important in health care social work as it is in criminal justice or child and family social work.

REFLECTIVE QUESTIONS/EXERCISES

How is compassion different from pity?
Discuss practice examples contrasting compassion and pity.
What is the role of empathy in compassion?
What are the obstacles to practising compassion? (For example, consider the issue of blame in response to human need.)

Compassion in practice

In Chapter 2 we noted that 'controlled emotional involvement' has long been regarded as a way of social workers orienting their empathic responses to the plights of others. This can be seen clearly as a crucial element of compassion, in that it finds the mid-point between pity and indifference. In practice, therefore, we ought to be able to identify the way in which this virtue is understood and expressed by social workers in reflecting on the way they engage with their work.

Let us consider the example of Cheryl, who works in a government-funded community agency that provides a broad range of services for children and young people. (As in previous chapters, this is a hypothetical situation, compiled from experience and observation.)

> Cheryl has been working for the last year with a young woman, Kristy, who has just withdrawn from a technical college course that provides training to work in social care and human services. Kristy tells Cheryl that the course was 'a bit boring.' Kristy is now at risk of losing her income as she is no longer entitled to receive the training allowance from the government social security department and she is not eligible for unemployment benefit because she chose to leave the course. Cheryl is also concerned that Kristy is not looking for work or an alternative course because she 'doesn't know what she wants to do with her life.' However, Cheryl notes that Kristy still wants to meet with her to discuss her situation, so she decides that this provides an opportunity to help Kristy look at the implications of the choices she is making and to consider alternative options. Cheryl asks the administrative assistant in the agency's office to book a meeting room so she can talk with Kristy, and her colleague says, 'After the work you put into helping her get on that course, I don't know why we bother some days.' Cheryl responds, 'That's okay. If she could sort everything out for herself she wouldn't need us.'

In this situation we can see that from some points of view Kristy has caused her own immediate problem by her choice to leave her college course in the way that she has: without discussing the decision, considering alternative programmes of study or looking at why she is reacting to the course in that way. She could, for example, have found another course at the college before withdrawing from the one she is currently enrolled in; or she could have recognised that she needs to talk about these things to make good decisions and have spoken with Cheryl first about other options such as looking for work. However, Cheryl sees in this a plight that many young people might face, especially if, like Kristy, they have grown up without having learned what others might regard as basic life skills. Cheryl's compassion avoids pity, because she is not seeking to deny Kristy's responsibilities in making decisions about her own life; rather, she recognises that Kristy's background means that even as a young adult Kristy still needs to learn to take responsibility. Cheryl is also committed to the view that when circumstances make it hard for someone to exercise agency, providing assistance does not in itself dehumanise. For her, an excessive response to Kristy's plight would be to take over and make her decisions for her. However, to say that Kristy had her chance and so she just has to live with the negative outcome of her decision is to go to the other extreme and to be indifferent.

Practitioners express their sense of compassion in different ways. For example, in one anonymised statement a social worker commenting on compassion observed:

> [Service users] want to see that you're like them. They don't want to see that you're a better person or you're just a technician [...] they want to see your humane-ness [*sic*]. They want to see that you'll stand with them side by side.

As well as this expression of genuineness, the term first used by Truax and Carkhuff (1967), we can also find an articulation of what they meant by warmth, in a practitioner's comment that compassion is 'being really present for someone'. Another practitioner put it slightly differently:

> If you can be someone that people look back on and [say] 'that person listened to me, believed me, I was able to trust that person, to disclose to them.'

Yet another notes:

> You can have all the models and theories that you like [but] if I can find that space of compassion in me, with the clients, then that's magical.

Routinely in the interviews that partly inform this discussion, when practitioners referred to compassion they linked it to empathy and understanding, to listening and to seeing someone's struggles with their problems. It is the personal quality that enables social workers to show empathy and to remain committed to seeking good outcomes for service users, even when those people react negatively. Social workers often encounter situations in which a service user may appear to throw away an opportunity (for a college course, for housing, for medical treatment and so on) because they are experiencing negative emotions (such as anger, despair or humiliation). This is where the capacity not to seek to ascribe blame enables social workers to maintain their empathic connection with service users. As one practitioner expressed it:

> that's where this system can be very difficult to work with, [and not to] fall into a blame-the-victim mentality, 'Oh it's all their bloody fault; oh they should just get their act together' [...] working with clients can be very frustrating because they often engage in self-defeating behaviour, but I think once you get into that victim[-blaming] mentality without a broader way of looking at things, I think that's a death knell.

Or, in the words of another social worker:

> I think it is just really understanding, not judging, those situations where maybe it's not ideal but that's the realities of [the person's] life.

Compassion enables social workers to be inclined to continue to offer assistance in the face of such apparent rejection or self-infliction of problems, through the combination of a sufficient degree of emotional involvement and the inclination to provide help (Halifax 2012).

However, there is also a challenge to compassion when social workers' roles require them to focus on what is not acceptable in the choices and actions of service users. For example, this can be encountered in various aspects of inter-personal violence and abuse. Child protection and family violence practice requires social workers to address the unacceptability of child abuse and domestic violence. This creates a tension for practitioners that is summarised by one social worker.

> What is most challenging for [...] this particular job, in domestic violence working with perpetrators [is] a tension between I guess the orthodox training [where] you walk alongside them, you empathise [...] and also challenging them and pointing out that what they are doing is wrong [...] being able to challenge them in a way that's respectful and productive that's not further oppressing them.

This practitioner adds later that he 'feels their shame' but at the same time understands that violence must be addressed in order to assist people to achieve change. (It is not clear if or how his compassion and empathy might be affected, as a male social worker, through gender identification.)

Social workers also recognise that there is a risk of the quality of compassion becoming affected negatively through the emotional engagement with work.

> I found this compassion fatigue thing – to go home when I've been running on adrenalin all day and then to put on a happy face and play with my son and have a conversation with my partner [...] it becomes draining [...] I think sometimes I've talked too much all day, I've dealt with everyone all day, I just want to go home and I've got nothing left.

This experience is shared with other social workers and is understood as 'compassion fatigue'. More widely, the demands of compassion are regarded as part of the phenomenon of 'burnout', which is discussed in more depth in Chapter 8 as part of the virtue of temperance. Specific to compassion, however, is that social workers appear to be conscious that although they feel this quality is innate, at the same time it can be enhanced or diminished through the experience of emotional exhaustion. This reaction comes from being able to see the common humanity in people's difficulties and being inclined to assist others while recognising that at least partly their actions may contribute to their problems, even though not being able to find a way out of their predicament without assistance is part of the reason why, for some people, this is a plight and not simply an ordinary life event.

Compassion is also regarded as a characteristic that can be gained or improved.

> I think I've got more capacity to encompass compassion for difficult people now [...] it's not that I didn't have compassion, but I've probably grown wiser in that regard [...] now I see mental health much more compassionately [...] more knowledge and more awareness.

Similarly, from a different social worker comes the statement that compassion can be enhanced purposively.

> At the moment I'm involved in a programme that's about compassionate care [...] it's a secular mindfulness thing.

As a personal quality compassion can at least be improved by training, if not actually acquired, not only through the formal continuing professional education of the kind described here, but also in supervision. Others have the view that compassion is simply more evident than before; for one practitioner, the overall approach of social work is more sophisticated than when she first trained.

> I think [there's] a lot more compassion now, a lot more grey areas than there were when I was 25.

In this reflection, the social worker is referring to the cumulative effect of changing approaches in qualifying programmes, ongoing training, supervision and the wider culture of the profession. This comment also indicates the way in which compassion is a response to increasingly complex theories of human need and practices that have developed to deal with such challenging issues.

So, compassion can be increased experientially, through the day-to-day experience of social work, and also in more formal ways through conscious attention. It can also diminish through fatigue, facing the apparent intractability of some issues, and less obviously through being gradually eroded by working in organisational cultures that do not value the characteristics of being able to stand alongside someone in their difficulties or distress, while not ascribing blame and not rejecting them because of disgust or anger.

REFLECTIVE QUESTIONS/EXERCISES

Can a social worker's compassion increase or decrease?
Why might compassion increase or decrease? (For example, consider the issue of 'compassion fatigue'.)
Reflect on how you might sustain or increase compassion in your practice.

The virtue of care

As Banks and Gallagher note (2009: 101–102), care is not a virtue that was recognised by either Aristotle or Aquinas. In both the ancient Greek and medieval European Christian world-views, the virtue of compassion alone is the basis for virtuous concern for others. Moreover, emotions were largely absent from this understanding of ethics and values, which were grounded in rational thought.

However, from the early modern period attention began to be paid to the role of emotion in ethics (which MacIntyre [1989: 162–170] associates largely with the work of Scottish philosopher Hume [1966/1751]). This change opens up the possibility of considering the way we feel has some bearing on what we value, what we think is good and right. For example, if I find that something makes me angry, or disgusted, or ashamed, then this reveals what I think is important and how I think things ought to be, what I care about (Nussbaum 2001). Emotion is not sufficient to guide the way in which I might then respond, or how I explain my values to others, but it guides me in recognising what my values are.

Slote (2001, 2007) argues that care may be considered as a cardinal virtue, one (even *the* one) that guides and organises the other virtues. His reason for this is that it is the only inherently outward-focused or other-directed virtue that is necessarily concerned with social relationships. As ethics concerns our orientation to the needs and interests of others as well as ourselves, the idea of a person caring, or being careful, as a core moral quality in itself appears logical. However, this might simply mean that the person has the characteristic of being considerate to what is rationally understandable. Instead, for Slote (2001, 2007) unless we also appreciate the way in which emotion plays a significant role, we will not have grasped the significance of care (compare with Oakley 1992; Blum 1994; van Hooft 1995; Sevenhuijsen 1997). From this perspective, it is emotion that gives care its moral content. While we may wish to allow that the connection between emotion and rationality is more dynamic than this implies, with neither having primacy over the other, it is certainly the case that without considering emotion we will not have addressed that part of what it is to be human that motivates our responses to the world around us (Slote 2001).

Care, then, is the virtue of being able to grasp and respond to those things that matter in the world. This might be termed 'being attentive'. Considering this from a feminist perspective has led to another ethical approach that also is grounded in the idea of care, that is, the 'ethics of care' (Gilligan 1982; Noddings 1984; Held 1993; Tronto 1993; Sevenhuijsen 1997). This approach focuses on the care of relationships as the core of ethical concerns, against the dominant approaches of rational attention to abstract principles (as in deontology or Utilitarianism). In place of such principles, Tronto (1993) argues for an ethics of care to have certain dimensions, namely attentiveness, responsibility, competence, responsiveness, integration of care. Although these dimensions have been considered as virtues in some discussions, we agree with Banks and Gallagher (2009: 102) that this is a misreading. The ethics of care focuses primarily on the person receiving care and the actions of the person caring in relationship to the one cared for. Indeed, the dimension of responsiveness refers to the reaction of the person who is cared for in response to the one caring, and the way the person who is caring receives this. In contrast, virtue ethics is centrally concerned with the character of the person acting. In this sense, if we consider attentiveness, responsibility or integrity as virtues, it may only be as personal qualities. From a virtue ethics approach, these are virtues because they are the qualities of a good person; the person is not seen as good simply because they act in this way. However, their meaning changes, even if only subtly, so that in all respects the virtues are attached to the one who cares, irrespective of the responses of others.

In the following discussion we follow Banks and Gallagher on this point. Although the ethics of care has provided a very important contribution to contemporary ethics and is certainly applicable to social work, even if it has not been addressed widely (Hugman 2005a; Barnes 2012; Bozalek 2013; Hay 2019), this is not our primary concern here. Care as a virtue is even less well considered in social work, despite the resonance that it has with social work as a 'caring profession' (Hugman 1991, 2005a). As a virtue, care may be understood as a disposition to attend to the needs and interests of others, to be responsive to those who cannot resolve their problems without assistance (England 1983: 13). In a detailed study of practice in Australia, Hay (2019: 368) records that social workers explain care in terms of 'meeting the clients' needs', 'just being there for clients', 'building relationships with clients' and 'going the extra mile'. Thus, they see care not only as relational but also involving a willingness to see and respond to what is necessary to assist the other person.

Feminist theories have also argued for the integration of disposition and action. For example, Finch and Groves (1980) and Ungerson (1987) write about the distinction between caring 'about' (which may be equated to attentiveness) and caring 'for' (which may be equated with responsibility). In this analysis care as a disposition cannot be separated from care as a practice (compare with Held 1993; Kittay 1999). This may weaken the claim for care to be thought of as a virtue, posing the question as to what is good about it if it does not lead to realisation in practice. The disposition must be linked to action and cannot be delegated to others. For example, at the centre of this feminist critique is an observation that having felt care 'about' a person or situation, the classic (and historically masculine) professions then gave instructions to others to actually undertake responsive action, in care expressed 'for' someone practically. Such delegation does not complete Tronto's element of the integration of care. MacIntyre's (1985) insistence that virtues are associated with practices provides a further basis for this view. Although a virtue can be understood as the dispositions of a good person, this ought not to be separated from the goodness of the actions that follow. It is not only that a good action is what would be done by a good person (although that is the case); it is also that a good person is disposed not only to feel responsible through being attentive but is disposed to do something about it – to act. On this basis, we can examine care as it appears in the qualities of social workers.

REFLECTIVE QUESTIONS/EXERCISES

What are your views about care as a virtue?

Tronto's (1993) dimensions of care are attentiveness, responsibility, competence and responsiveness. Where do you see these dimensions in your practice?

How does care relate to compassion?

Care in practice

Care as a virtue can be seen in the following hypothetical situation in social work in a hospital.

> Dionne is a senior social worker in a sexual assault service based in a large general hospital. She recently had to intervene when a young woman, Sandra, who was leaving a medical ward (having been medically discharged), came to the sexual assault service to ask for help, became distressed and then started to be verbally abusive and threatening towards a younger, less experienced colleague, Sophia. Dionne had overheard Sandra starting to explain her situation in a loud voice, and had noticed that she looked distraught as she entered the social work office. So when Sandra's voice became so loud that Dionne could hear her through the office wall, Dionne knocked and just entered the interview room. She found Sandra standing in a threatening manner with her fists clenched, appearing quite aggressive towards Sophia. Dionne recalled that Sandra had had a long experience of physical violence from her male partner. From this observation Dionne realised that Sandra's actions were not because of anything Sophia had said or done, but that due to her trauma Sandra had been having flashbacks the previous night, after being told she would go home the following day. So Dionne focused on helping Sandra to calm down enough for a plan to access supported accommodation to be made. Later Dionne debriefed with Sophia and helped her colleague to understand Sandra's feelings, while also herself explicitly focusing on Sophia's reactions to the aggression and how she felt about it.

In this encounter, Dionne reveals a strong quality of caring. She responds to the challenge of aggression, and her colleague's experience of being threatened, by being concerned for the wellbeing of both the service user and the social worker. Dionne's disposition is to consider the needs of both and to take responsibility for responding in ways that enable both to feel recognised and attended to. Care here is seen in the way that Dionne is inclined to be concerned about each of them and then to care for them in the ways that she acts. Care comes from Dionne's innate emotional response, which is then expressed in practice.

Recalling a particular situation in which she was seeking to be attentive to a young man, a health social worker interviewed by the present authors noted that she was at some time told she was the type of person who appeared to be caring.

> [Service users sometimes] say 'You've got a kind face' […] the other day, this really standoffish 22-year-old boy [*sic*], he gave me a bear hug, you know, that brings up other issues, but it was really lovely. It was the middle of the night and his mother was dying and he had been standoffish for the last six hours and I'd just given him space, and not intruded, and come back and I was just saying goodbye and suddenly he stood up and went like this [gestures a hug].

Although this instance has some similarities with Sevenhuijsen's (1997) description of an interaction in a nursing home for older people, whereas Sevenhuijsen says this is a demonstration of the ethics of care, if we focus on the social worker who is expressing care we have a clear view that this is an ethical quality that she possesses. Like Dionne, this social worker's character leads to action – that is, what a virtuous person would do; it is not the act that gives her the quality of character. A different social worker makes this point very simply when he says:

> [Service users] want to see kindness. They want to see empathy.

While this may appear to be simplistic, whether knowingly or not it repeats the findings of Truax and Carkhuff's (1967) analysis of the views of service users about the qualities of effective social workers: warmth, genuineness and empathy.

At the same time, we should note that in the account above of Dionne's responses to the particular needs of a specific service user and her colleague, there is evidence of the use of authority. This introduces the difficult question of 'control', which is often understood in social work as antithetical to care (Day 1981; Howe 2014). Dionne acts firmly to intervene to prevent harm, including the harm that the service user's actions might have caused to her own self as well as to the other social worker. In this Dionne displays the technical quality of decisiveness, which is allied to her ethical quality of care that inclines her to seek the good for both the others who are involved.

Bauman (1993: 92) observes that '[t]here is but a thin line between care and oppression'. Tronto recognises this in what she calls the 'dangers' of care, which are seen in maternalism or paternalism and in partiality (1993: 170). By this she means an inequality of relationship that can lead a caregiver to value their own competence over that of the person receiving care, or to favour those people or principles that the caregiver holds as valuable, irrespective of who the cared-for person is or what she or he values. Another danger is when the caregiver expresses feelings of condescension (which is comparable to pity in relation to compassion). The caregiver may actually be more competent; what matters is that the person receiving care is able to express their own needs or to reject the care that is offered and not to feel they are 'being looked down on'. The question about 'control' in social work comes from limits to the capacity of those receiving care to assert their own perspective. This usually is grounded in the statutory framework of the practice. Central to this issue is whether control (the exercise of power) invalidates social work's claims to other values, such as respect, dignity, rights and justice. In this particular instance, we might reasonably regard the idea that allowing the service user to express her feelings through aggression and then face the consequences as the achievement of respect for her dignity as a human agent fails to balance these values with the ongoing implications for her – not to mention the reasonable need for safety for the social worker (compare with Gilligan 1982: 28). Care suggests that there is nothing dignified, respectful or just in standing back and watching someone hurt her or himself or someone else when intervention was possible.

In this sense, control in the form of setting limits may be a means to express care and so to be consistent with Dionne's inherent care. However, Bauman's caution about the fine distinction between care and oppression noted above points to the impossibility of devising a principle that would tell Dionne how to act (Bauman 1993: 91). Should she impose her own view of another's wellbeing, and so override her human subjectivity, or choose not to act to avoid taking away the other's autonomy and so display indifference? For Dionne in this situation, because of the moral scale in the possibility that Sandra might have harmed Sophia, to choose not to act would have placed a greater value on abstract principles than on the humanity of either the person who is driven by harmful emotions (Sandra) or the other who is threatened (Sophia). Thus, she chooses to intervene, but to do so in a way that respects Sandra's subjectivity and gives her opportunities to address her problems and issues.

What then of different social control functions, such as in child protection or criminal justice (probation/corrections or prison welfare)? Is care possible as an ethical virtue in such practices? Or, from a virtue ethics perspective, is it possible for a caring person to practise in social control contexts? The answer to these questions is that the ethical quality of care is about the innate disposition of the social worker; it is not about the nature of the institutions in which practice occurs. We might agree that for some social workers who have the ethical virtue of care, such a setting may be unconducive to their professional approach. Others might well find that they can bring the quality of care to bear on tasks that are necessary, given the wider social context. However, what is required in such work is that care is joined with the ethical virtues of compassion, respect, dignity, humility and justice (as discussed in other parts of this book). Caution about authoritarianism must be added to Bauman's (1993) and Tronto's (1993) concerns about the risks of care in maternalism, paternalism and partiality.

The practices that support the quality of care in social work have been identified above as warmth, empathy and genuineness. These in turn require the skill of being able to listen 'carefully' (in both senses of this term, with accuracy and sensitivity). Can we then consider these as elements of the ethical virtue of care? To a large extent this appears to be the case, as these qualities can be understood in practice. However, these are also ethical characteristics that care shares with compassion. Therefore, it may be more helpful to consider the ways in which these two ethical virtues are linked and support each other.

REFLECTIVE QUESTIONS/EXERCISES

Consider the actions of the social workers integrated as case examples above – how do you think you would act in such situations?

What do you think are the critical issues or complexities in caring?

Think about practice situations where care and control need to be balanced – what issues does this challenge raise for you?

Connecting compassion and care

The common element that is identified between compassion and care is the capacity of empathy. In many discussions in social work, empathy is referred to as a quality in the practitioner. However, insofar as this is correct, it might be considered a 'technical' virtue (see Chapter 1) – empathy is not in itself an ethical characteristic. Yet, there are two reasons why empathy can be seen as important for these ethical qualities. First, the way in which it is used by social workers is the concern of ethics, as empathy can be used for bad as well as good purposes. For example, the effective torturer must be able to grasp the emotional world of their victim, emotionally as well as intellectually, in order to terrify, which is a form of empathy. It is where empathy is harnessed to compassion and care that it enables social workers to pursue good objectives and in itself becomes a basis for good practice. Second, conversely, compassion and care require empathy in order to be reflected in practice. Otherwise, as dispositions they may become ineffective and have no impact on the social worker's actions.

Compassion and care also support each other. For example, in counselling a man who has deliberately physically harmed his child, a social worker who is disposed to care for him must also have compassion, perhaps overcoming anger and disgust at the objective nature of the man's actions. Indeed, it is difficult to envisage how effective social work of this type can be done unless practitioners have the qualities of compassion and care. Previously, we noted that Slote (2001, 2007) regards care as the central ethical virtue, which sustains all the other virtues. We do not have to agree with this assertion of its primacy to concur that, together with compassion, it is vital for good social work. The practitioner who is disposed to care may still feel a lack of compassion for some people or circumstances (such as the father who has abused his child), but as a response to her or his inclination of care chooses not to work in a particular field of social work (in this instance, child protection).

For these reasons, compassion and care appear to be very closely connected, perhaps even as two sides of the same moral quality. To some degree it is difficult to envisage social workers who lack these characteristics. Even in structural or macro practice, compassion and care are allied with justice, respect, dignity and rights as powerful motivators. In difficult and challenging social work of whatever mode, unless the practitioner has a strong innate sense that it 'matters', in other words feels compassion and care for those who they seek to assist, as we have understood these qualities here, social work is likely to be mechanistic and even ineffective. The importance of compassion and care as significant ethical virtues does not devalue justice, respect, dignity and rights; rather, it adds to our understanding of the commitment needed to undertake this work.

REFLECTIVE QUESTIONS/EXERCISES

What are your thoughts on how compassion and care are interconnected in practice?

Can caring be possible without compassion?

Cultivating compassion and care

As we noted in the previous chapter, because they are aspects of character, ethical virtues may appear simply to be innate, in that a person has or doesn't have them, or may have them to greater or lesser degrees. Yet, as with courage, compassion and care can be developed and improved. It is highly likely that (at least most) people who enter social work do so because they have these virtues to some degree (Banks and Gallagher 2009: 110). At least in most countries, social work does not have a high-status appeal that might attract entrants for other reasons (such as financial reward).

Yet, it is also important that social workers become aware of the problems and pitfalls, including maternalism and paternalism, partiality and condescension. Indeed, without such attention and 'self-care' social workers can run the risk of burnout. Self-awareness of this kind can be innate or it may be developed through encouragement, for example, in supervision. Banks and Gallagher (2009: 109) also note that these qualities can be enhanced by experience, such as through role plays and discussion of situations taken from actual practice in formal training, as well as the modelling of good practice while on field education placements (practicums).

Moral exemplars are a powerful source of how we might understand ethical virtues and consider them for our own development. However, we cannot rely on such colleagues being available for every new entrant to social work, or for those who are at later stages in their careers but still need to develop in this way. For this reason, formal training programmes could play a more conscious role in addressing these personal qualities, while for those who are further advanced in the profession explicit attention in supervision might be considered.

Conclusions on compassion and care

In this chapter we have examined compassion and care as ethical virtues. Both arise from the emotional engagement of practitioners with those who receive social work services and with the issues and problems that they face. Although the meaning of each is different, compassion and care are closely aligned and support each other. Both are positioned between indifference (to people and to their problems) on one side and to an excess on the other (in which the compassionate or caring person has a sense of moral superiority to the one who is in need). Yet, without either of these qualities it is difficult to envisage what might bring someone into social work or, perhaps more importantly, hold them in what is a demanding and at times under-valued profession. Empathy is a capacity that enables practitioners to act on the virtues of compassion and care, but without those dispositions empathy in itself lacks a moral direction (and can be used for bad as well as good purposes). Compassion and care are not binary, in that they are not either present or absent, but may be experienced in degrees. The challenge for the good social worker is how they may be nurtured and sustained, in order to maintain commitment and good engagement with those who receive social work.

FINAL REFLECTIVE QUESTIONS/EXERCISES

What have you learned from this chapter? List three main aspects of learning.

How can we further strengthen the qualities of compassion and care? How might you do this for yourself in your own practice?

5

HOPE, PERSEVERANCE AND RESILIENCE

> **CHAPTER OBJECTIVES**
>
> - To define the virtues of perseverance, hope and resilience, and explore how they relate to and support one another
> - To explore examples of hope, resilience and perseverance across different social work practice contexts
> - To consider how hope, perseverance and resilience may be cultivated in the social work profession

The virtues of 'keeping going'

Social work often requires practitioners to maintain a positive view about possibilities for change and a commitment to long-term work in which progress may be slow or there may be setbacks. In this chapter we look at three interconnected virtues that contribute to the potential for social workers to be able to 'keep going'. These are hope, resilience and perseverance. Although they are distinct and we discuss each one separately, they support each other, so we also consider the way in which they intersect and are often woven together.

Social work often involves intervening with people in the social issues and problems they face when the solutions are neither quick nor easy. Yet qualities such as hope and perseverance are seldom addressed in social work ethics. Resilience has become more widely recognised as important in social work (as well as in other areas of life). For the most part attention to resilience has been to consider it as a psychological mechanism and to concentrate on the ways in which people can develop this as a skill. In that sense, resilience can be understood as a technical virtue (as explained in Chapter 1). Likewise, although Flesakar and Larsen state that hope is an orientation (2012: 72), they emphasise that it is a skill to be learned (2012: 74). However, in this analysis we examine it as a moral quality that, together with resilience and perseverance, contributes to the disposition of social workers to 'keep going'.

Another term for this overall group of virtues might be 'tenacity'. This refers to the commitment not to let go of something. From an Aristotelian perspective, tenacity, and each of the more distinct qualities of hope, resilience and perseverance, have associated 'vices' of excess as well as deficit. It is possible to be too tenacious – that is, overly hopeful or to persevere when changing direction would have been wiser – just as it is possible to give up or give in too easily. There is an issue in practice of the risk of being too optimistic. This can be seen as an ethical problem because it can lead to professionals encouraging service users to take risks that are not well supported by evidence, without making clear that the course of action is risky. However, Flaskas (2007) implicitly suggests that a greater challenge lies in failing to be hopeful enough, leading to a lack of perseverance. Whether it is possible to be overly resilient is a different question.

At this point it will enable us to examine these different qualities more thoroughly if we separate them. So, we now turn to each of hope, perseverance and resilience in order to look at how they contribute to good social work practice when we understand them as ethical virtues.

Hope

Like compassion and care, which we addressed in Chapter 4, hope has a very clear emotional dimension (Oakley 1992). In everyday life it can be regarded in terms of 'having an optimistic nature'. Thus hope can be explained as an innate disposition to grasp the possibility that future events will be good, or that problems can and will be resolved. However, like compassion and care, hope is not only an emotional response in and of itself, but is better seen as a disposition that has an emotional dimension while also containing a cognitive element and, for some social workers, also a spiritual component (Koenig and Spano 2007). For example, the American Christian progressive social campaigner, Jim Wallis, is widely cited as stating that hope is 'believing in spite of the evidence, then watching the evidence change' (Loeb 2004: 22).

However, this appears to open up the possibility that hope is naïve and overly optimistic. Indeed, if we accept that it involves a tendency to assume that good outcomes will be achieved from a combination of good intentions and perseverance, even if there is overwhelming evidence to the contrary, then this could be taken as an excess of this quality. Yet, both Wallis and Loeb have long experience of working for social change, including with others such as Desmond Tutu and Vaclav Havel (Loeb 2004; also, see Tutu 1999). There is a comparison also with other forms of community development. Thus there could be said to be a type of evidence that supports optimism, namely that change may take a long time and be difficult to achieve but bad situations can improve, for example, as argued by advocates of environmental or 'green' social work (Hoff and Polack 1993; Besthorn

2002; Dominelli 2012; Alston 2013). Hope is what inspires people to persist and to continue to work towards what they consider to be good.

Loeb (2004: 22) argues that hope is not simply wishful thinking; rather, it is a way of living, which he also connects with perseverance and resilience. This appears to render hope as an act. The hopeful person in this way of thinking is the one who acts on the basis that future events will eventually be good – the action of 'hoping' follows from the quality of hope. To correct this, Flesakar and Larsen (2012: 62), citing Havel (2004: 82), refine this definition, saying that it is not a matter of knowledge that events are going well or even an expectation that they will in the future, but rather the inclination to work towards something because it is good in itself, not depending on whether it will be successful.

In counselling and family therapy practice, Flaskas (2007) argues that hope is a crucial dimension that enables practitioners to focus their interventions towards good outcomes. She refers to counsellors and therapists as having 'strong personal reservoirs of hope' (Flaskas 2007: 195). This implies that it is the innate qualities they bring to their practice that enable them to find the possibility of hope for those they seek to assist. At the same time, Flaskas (2007: 191) argues against what she regards as a binary division between hope and hopelessness. Instead, she proposes that the capacity to hold them in an appropriate balance can, in fact, assist people to cultivate perseverance and resilience. In recognising that hopelessness remains as a possibility, the hopeful person maintains a grasp on the challenges against which hope is held. However, conscious attention is required in order to avoid an imbalance of over-optimism or despair (Flaskas 2007: 196). What she means here is that the disposition to hold onto the possibility of what is good should not deny the reality of what must be faced and changed in order to achieve that. In other words, hope must not be naïve or simplistic, but may even require other virtues such as courage and wisdom to be sustained.

Flaskas' (2007) example of the family therapist having hope that change is possible and bringing this into their practice points to one mode of practice in which this characteristic has relevance. Loeb (2004) and others cited previously demonstrate that hope is also significant in community practice. So, we can say that hope is a characteristic of the good social worker across different modes, from micro to macro levels. All social work practices are directed towards assisting others to achieve change, whether it is in themselves and their personal relationships or in the social structures and systems that comprise the world in which they live. To remain focused on goals when the obstacles are objectively 'large' requires a perception on the part of the practitioner that change can be achieved. Consequently, the search for an evidence base that will guide practice to what is most likely to be successful is logical (Smith 2004). At the same time, the objects of social work practice (such as personal relationships and social systems) are highly complex and indeterminate. Indeed, we might say that the range of variables is so great that the degree of predictability of the natural sciences is impossible to replicate (although the natural sciences actually often also address probabilities rather than certainties – see Hugman 2005b). Social work frequently operates on the basis of probabilities, possibilities and potentials. For this reason, as a personal quality hope has much to offer, as it provides a sustaining orientation towards complex practice.

Hope in practice

One aspect of hope that bridges micro and macro modes of practice can be identified in strengths-based approaches. Originally developed in community practice (Saleebey 2002, 2004), the strengths-based approach has also had an impact in micro practices (Corcoran and Pillay 2009). For example, a social worker in mental health counselling told us:

> that's what we're here for [...] to recognise some hope in some really awful scenarios, the silver lining to things and working towards [that]. So even though they might seem unattainable to yourself and the client, that they're worth working towards.

Another who works in child protection commented:

> I don't believe in dead ends [...] If that didn't work, what else can we do? Keep trying.

So, in these micro practices hope is understood in terms of being able to keep sight of the goal of intervention, rather than focusing on the apparent intractability of problems. It equates to maintaining a sense of optimism within the self that provides the basis for encouraging service users to remain engaged with the process of counselling or family therapy.

Similarly, a social worker in community practice observed:

> there's always that hope. There's always something you can find to empower [...] some strength that you can find in people [...] so I work from a strengths perspective [...] always look at what the strengths are in people.

For a different practitioner, hope is found in one's self, rather than in the interaction with others.

> It was uncertain whether community members would attend. I was telling Council that it would work, but I didn't actually know if community members were going to attend [...] there could have been something that happened in the community and nobody would've come along.

Another summarises the insights they have gained from their career as:

> We bounce up. Every time we think that bar can't be lowered any more, it is [...] but we constantly bounce ourselves back up and try again and try

> again […] I would say 'always keep hold of hope.' That's one of the big things. And I would say 'keep fighting.'

In this sense, hope is a disposition to remain committed to working to change obstacles to achieving the goals of the community and its members.

Throughout social work, practice is focused on achieving change (even if, as we noted in Chapter 3, such change is to prevent decline and maintain a satisfactory life). However, hope is experienced within the person and seen by others in response to obstacles. Some of these come from within the self. For example, hope can be challenged by fatigue, whether this is physical, psychological, emotional or spiritual. Social workers are aware of 'compassion fatigue' and its contribution to 'burnout' (Pines and Aronson 1983; Kiley at al. 2018). However, self-care here includes the Socratic sense of 'caring for one's moral self'. Finding a balance within the demands of vicariously engaging with others' problems and the individual and social structural causes of those problems must be recognised explicitly and addressed. We return to this point about balance in Chapter 8. Here the point is that hope can be undermined by the demands of practice and must actively be sustained.

One way in which the obstacles encountered by social workers that challenge hope can be seen is in the difficulties service users have in responding to the hope of the practitioner. Sometimes this is because people become 'stuck', in the sense this term is used in therapeutic work (Flaskas 2007: 192) to refer to being psychologically immobilised by the effort of seeking change. It might also be because of the impact of responding empathically to others' pain leading to forms of vicarious trauma (Pines and Aronson 1983). Trauma has a variety of origins, including abusive family relationships and also the ways in which social systems operate. A particular instance of this is in the experience of refugees and asylum seekers, who, having been subject to physical violence and the psychological harm of being forced to flee from their homes, are then subject to complex systems in countries of settlement. Health and welfare policies and the ways these are administered may cause further trauma. A practitioner in this field said:

> in a crisis you're not doing transformational work, you're actually most of the time doing transactional work […] but if that dominates all your work it is not satisfying for the client and they disengage, but it's not satisfying for [people who] became social workers because they were interested in transformation and flourishing and hope and resilience.

This person went on to say that the biggest such obstacle is government policy. This can be in the form of the way policies create the problems for service users, or it can be in the impact of policies on services themselves, such as regulations or funding.

There are various ways in which social workers find hope or seek to maintain it when they recognise that it is challenged. First, some find it in a broad sense of the ethical values that brought them into social work originally. Retaining a strong commitment to these values becomes a source of a reason to think or believe that change can be achieved, in the way that Loeb (2004) identifies, as we noted earlier in this chapter. This way of sustaining hope might also be seen as a commitment to the profession of social work and its wider goals. Another possible foundation for this source of values is political ideals.

Second, as a personal quality, hope can be buttressed by spirituality or religious faith (Koenig and Spano 2007). Faith of this kind also can be a motivating factor in someone choosing social work as a professional career. As with commitment to the values of the profession, religious or spiritual faith contribute to a practitioner's sense of self and the qualities that they bring to their work. Both can be referred to in terms of 'belief', which echoes the sense of conviction that good outcomes will be achieved by pursuing good objectives. Such beliefs find their strength in recognition of something greater than the individual self, a source of meaning that can be understood as external to the person.

Whether or not this is regarded in religious or spiritual terms, or as a broader commitment to values or political ideals, the important common element is that the person finds the basis for hope in something other than her or himself. Indeed, for some social workers, religious or spiritual faith and political or ethical convictions are not separate, but are different facets of the same qualities of character. This in turn may provide those who have both strong religious and political commitments that are held together to find hope in this combination. Although there can be reasonable concerns about the way in which social workers with strong religious, spiritual or political values may impose these values on service users, whether deliberately or inadvertently (Clifford & Burke 2009), at the same time these are powerful positive forces to sustain practitioners in the face of the demands of social work. Others do not need to accept the basis for a social worker's hope in order to gain from it and be encouraged. As Olver puts it, hope allows 'different perspectives to co-exist rather than having to reach consensus' (2012: 240). When held in this way, hope enables both practitioners and the service users to find a reason to keep trying. In this way, hope is very closely aligned with the virtue of perseverance.

REFLECTIVE QUESTIONS/EXERCISES

From where do you draw hope?

Think of a time in your work that your hope was challenged or undermined. What were the challenges?

How do you sustain hope in the face of challenge?

Perseverance

In classical ethics, perseverance is often referred to as 'fortitude' (Pellegrino and Thomasma 1993). This can sound out of place in the early twenty-first century in contexts in which rapid change is accepted as normal. Yet changes in social relationships or social structures can take time and require those involved to remain committed to processes that are often neither linear nor smooth. Setbacks, obstacles and the pressure of competing demands are common in social work, so without an orientation to 'keeping going' there are many reasons why practitioners might find reasons not to pursue goals that are difficult to achieve.

Pellegrino and Thomasma (1993: 113) also relate perseverance to courage. It is necessary, they argue, for a practitioner to have sufficient moral strength to maintain their commitment against various challenges. Setbacks and obstacles can arise from service users being unable to maintain their contribution to the process, or from organisational decision-making and wider social policies (such as restrictions on budgets). In Chapter 3 we considered courage as an ethical quality that relates to social workers having the orientation to address all of these issues. However, without the corresponding virtue of perseverance, courage is difficult to maintain.

While perseverance is an ethical virtue from this perspective, it also carries with it the same risk of excess as we see in hope, namely knowing when to stop. The problems that come from the excess of perseverance are seen in consequences such as unintended harm that is caused to either the service user or the practitioner, such as from the reactions of others to what they perceive as an unreasonable demand for their response, using up resources that might have been directed to other needs, or from the less tangible but no less real possibility of fatigue or disillusionment of the service user, practitioner or both.

Beyond identification of it as a contributing factor to courage, relatively little has been written about perseverance as a quality of social workers. Two exceptions are in a discussion of a mental health project (Luntz 1991) and an analysis of child protection practice (Saini *et al.* 2012). First, in the mental health project, issues of setbacks and obstacles were seen both in the internal relationships of the team itself and in the structural environment in which it was being formed. Luntz (1991) describes how the processes within the team caused instability for the group, but that while they were resolving these there were changes in the external organisational and political environment that created additional stresses. Changes of key members of staff at critical stages then compounded the situation. As Luntz observes:

> Pioneering activity, in this case the creation of a new way of working and the development of new knowledge, is arduous and demanding. As the full-time staff members began developing clinical skills, the idealism that had drawn them towards preventative work began to wane.
>
> *(Luntz 1991: 44)*

The perseverance of key members of the team to pursue the realisation of these ideals is then attributed as a critical aspect of the way in which, over time, this project succeeded and began not only to provide more preventative care for families, but also the impact this had on practice and policy elsewhere.

In the example of child protection in situations of parents divorcing, Saini and others note that the social workers in a qualitative research sample provided evidence of the need for the capacity to cope with high levels of stress and 'trepidation' (2012: 1313–1314). Some of these feelings came from working with the volatile relationships within families, but also often from the demands of other professionals, including lawyers, teachers and other social workers. Perseverance in this situation is the disposition to maintain the quality of professional practice that is needed to produce good outcomes for the families concerned, especially the children. Moreover, in the same way that we have noted for the mental health project, here too perseverance is very closely interwoven with courage in the everyday demands of this practice. As with that project, perseverance also implies the strength of character to maintain professional commitment, expressed in the willingness to learn, to collaborate and to resolve specific problems.

These two examples provide accounts of perseverance as 'keeping on going'. It may be that instances of the limits of perseverance are not reported because it may feel like 'failure'. There is a sense that patterns of work (such as organisational structures and methods of intervention) change when they come to be perceived as no longer adequate. This certainly is noted as the intention behind the creation of the new mental health project (Luntz 1991: 42). However, because analyses of how practitioners might decide that something is not succeeding are not shared in the same way as successes, there is little evidence of how recognition of and a willingness to act on the limits of perseverance can be examined as a positive ethical quality. We return to this point in Chapter 7 in a discussion of the virtue of humility.

Perseverance in practice

The statements made by social workers to the present authors about the characteristic of perseverance are varied in both the scope of the practices that they address and the way in which they are understood to relate to other ethical qualities. For example, references were made to the length of time that it had taken to see progress in a service user's situation. One caseworker in community health services commented:

> When [this person] first came [to the service] she was referred to two or three different services and they fell away [withdrew] [...] when we engage we do all we can, so we will ring, try again, try, we don't give up easily [...] but after we've been involved for 12 months then they've all come back on board.

Another practitioner in community health in a different part of the country likewise noted:

> It took a long time, but that didn't worry me much because I still was accompanying her on the journey, if you like [...] just realising that all I had to do was be there, be there, be there. A long time of being there, we're talking five years [...] one of the big things I've learned here is that persistence, that sticking with people long term.

A social worker in palliative care was slightly more succinct. When asked about the important aspects of her practice she simply said, 'patience and consistency'. A child protection social worker in a different part of the country used the same terms, stating that a critical quality in developing good rapport and communication is 'persistence'. Yet another said:

> I think one of the biggest things I have learned here is that persistence, that sticking with people long term.

This sense of perseverance as persistence is also expressed as an explicit part of the interaction between social workers and service users. A practitioner in community mental health described perseverance as a way of interacting with service users who find difficulty in maintaining their own engagement even though they insist they are committed to seeking change.

> I made really clear to her at the start that I was not going to give up on her [...] she'd tried lots of things in the past [...] I said to her, 'This is going to be hard work and you're going to want to quit, and I'm not going to let you quit. I'm going to ring up and I'm going to push' [...] whereas sometimes [professionals] say, 'Okay. You don't want to come' [...] and I'm saying, 'I'm not going to do that. I'm going to [...] if you miss an appointment, I'll be ringing if you don't come in.' [Laughs] And I think that you have to [...] you work with [someone] [...] like, this particular client, you work with them, you are personally invested because you work with them for so long [...] probably going the extra yards [...] then when she had successes we celebrated those. Because it was tough.

For some social workers, perseverance complements hope. A health social worker connected the two virtues.

> Perseverance probably [...] we started in December and we finished about the middle of June [...] so probably perseverance and [...] maybe in a way it's about having faith in the systems, so just stick with it, stick at it.

Although many instances of this quality are described by those who are involved in micro-level practice, more macro practices also demonstrate this sense of perseverance. A refugee resettlement social worker recalled:

> [...] what I did well was that I didn't give up the fight for this family. That I was very clear in the information I was providing to [the Immigration Department], to him, to everyone about what was happening. That then enabled me to be present in those meetings which are very intimidating [...] to be very articulate [...] I just held this family very strongly in terms of trying to get them the services and the outcomes they needed.

Hope is connected with perseverance, whether or not in any one instance the goals are reached or the planned intervention encounters setbacks or obstacles. Some of the social workers quoted here reported that the work they were undertaking had not as yet achieved the overall goals that had been set. However, in these instances, the sense of optimism remains that without perseverance the outcomes for people who are suffering, are marginalised or are oppressed in some way can be much worse.

A major obstacle to perseverance is that social workers often practise in organisations that have insufficient resources to meet all the needs with which they are presented. Although resources are finite (whether on a global or a local scale, whether financial, material or the numbers of practitioners and the hours in each day), social workers often face questions of access to resources and their equitable distribution. We return to issues of justice and fairness in the next chapter, but at this point it is important to make the connection between such qualities and the virtue of perseverance. The quotations above include references to pursuing equitable access to services as a central aspect of social work and link this to the quality of perseverance as part of the way in which they are inclined to direct their work.

In other instances, perseverance on the part of the practitioner is evoked in the way in which some service users themselves lack the capacity to persevere in the difficult tasks required to achieve change. This is the case whether the change is in their own responses to the relationships and systems around them or changes in those relationships or systems. In such circumstances social workers who have the quality of perseverance may be able to offer this to those they seek to assist. Whether or not service users respond to this perseverance or not, the quality sustains those social workers' practice. As a school social worker put it:

> [one service user] has called me annoying many times, 'You're just annoying,' and I know what she means, 'and even if I don't want you you're there' [...] [but] when the dust settled [...] and everyone else went away, I didn't want to be one of those people [i.e. the sort of person who does not persevere].

What is important is that perseverance comes from the characteristic of 'sticking with' people and the difficulties they face.

REFLECTIVE QUESTIONS/EXERCISES

Describe the relationship between perseverance and courage. How might these virtues contribute to and sustain one another?

What are some of the 'risks' of an excess of perseverance?

Think of your own practice – how do you know when something is not succeeding, and it may be time to let go or move on?

Resilience

Theory and practice methods in social work have increasingly attended to the notion of resilience. As a concept, resilience is usually regarded as a psychological response that enables people to cope with challenges in their lives (Worsley 2015). For this reason, it tends also to be discussed in terms of assisting service users to develop psychological awareness and strategies of how they respond to such difficulties and so to develop resilience. As with the ethical virtue of hope that we discussed earlier, resilience understood in this way is also frequently associated with theories about strengths and the related 'strengths approach' to various modes of practice (Saleebey 2002, 2004; Ungar 2008; Worsley 2015).

In contrast, as an ethical virtue, resilience concerns the character of social workers themselves. In this sense, the capacity to withstand challenging or harmful life events is not only psychological but also has a moral dimension, in that it is a good in itself and so part of the ethical character of the social worker. We may describe the ethical quality of resilience as that aspect of character that disposes someone to withstand the moral stresses that come from life challenges and difficulties. In previous chapters we have argued that virtues can – indeed must – be cultivated. This implies not only that they can be developed but, conversely, if not sustained they can decline. Given the nature of the problems and issues that social workers address in their work, this is a quality that contributes to the maintenance of the moral self (self-care in the Socratic sense). In these conditions moral resilience is an important dimension of the overall ethical character that enables social workers to have hope and to persevere.

Understood in this way, resilience is also (together with perseverance) substantially an expression of the classic way of defining the virtue of fortitude (compare with Pellegrino and Thomasma 1993: 110–111). Yet it is not exactly the same. Whereas perseverance addresses the mid-point between giving up too easily and never giving up – even when wisdom suggests that is the better choice – resilience refers to that quality that enables a person to face the setbacks and obstacles that might cause someone to give up on something that is valued, without losing their grasp on their own moral self or being able to 'bounce back' after a setback.

Is it possible to think of having a deficiency or an excess of resilience? If we understand the resilient person as one who can withstand the moral as well as the psychological and emotional impacts of negative life experiences, having too little resilience clearly is a problem for people in general, not specifically for those in a profession such as social work. Being morally, as well as psychologically and emotionally, defeated by life events is a barrier to coping. However, this question raises a secondary query, which is whether there is a sense in which a degree of being hurt by such occurrences can actually play a positive part in someone's moral life. Given the way in which it is widely agreed that moral virtues are interconnected rather than discrete, we should examine the potential for an excess of resilience to be such a possibility. This then leads to considering whether a person who displays excess resilience is one who becomes insensitive to the problems of human living, or plights as we discussed in the previous chapter (Nussbaum 2001). In other words, it could be related to someone having too little compassion or care. At the same time, it is distinct in that what sustains resilience is the capacity to experience negative events in human life, whereas compassion and care have their origins in and are sustained by other factors.

We also should not see resilience as dependent on rationality. Although an accurate grasp of balancing positive aspects of human life – such as evidence of how people can resolve difficulties and the likelihood of this happening – does not of itself produce resilience, this may play a part alongside psychological and emotional factors. In this way resilience plays a role in developing the moral self, drawing on technical virtues, as discussed in Chapter 1.

Resilience in practice

When discussing resilience, social workers tend to focus on the way in which it is evident psychologically and emotionally in service users. This can be seen both in the literature on resilience (such as: Ungar 2008; Worsley 2015; Walsh 2016) and in the way in which the social workers with whom the present authors discussed ethical virtues talked about their work. However, in conversation some also identified resilience as a quality that they recognise in their own character. For example, a social worker in emergency management (disaster response) practice said:

> I think I've got personal resilience [...] you can't just chuck the towel in and go, 'Forget it, I think I'll pull the blanket up and go to bed.' So I think it's my own personal resilience.

Others convey a sense of resilience being developed and brought out by their practice experience. A child protection social worker said:

> I think I've developed immensely, not only through becoming a mother but just life experiences and being thrown into the deep end of child protection, I guess. I've seen some pretty wild things now, so I think that forces you to mature a little bit [...] and harden a little bit. But I don't really like the word 'harden,' it's not necessarily that. I think you build resilience.

For this practitioner resilience was something that had to be developed; it was not simply an innate quality from the start. This is the experience of others, too, such as the mental health practitioner who revealed:

> I really had to dig deep and I really had to find the resources within me to not be beaten [...] there was a point when I thought 'I don't know if I could do this, this is way too hard' [...] but I just kept going [...] and I'm really proud of that, because I've not always been [...] the most robust person [...] that really taught me a lot about myself.

This reflection illustrates the connection between perseverance and resilience, in the way the former contributed to the development of the latter.

A different understanding is provided by the practitioner who described awareness of her own resilience as a quality that could weaken as well as grow.

> I have a fair amount of resilience, and some of that has ebbed and waned over the years. There were times when I felt beleaguered and stressed and I would get help [...]. So resilience is important to me, compassion is important to me [...]

Here the sense is of compassion as a virtue that supports self-care as well as the care of others. While some of the stressors to which this practitioner is referring came from the vicarious impact of engaging with the problems and issues faced by service users, there was also the effect of organisational challenges, including the politics of large structures as well as the actions of senior colleagues. She subsequently linked resilience to avoiding burnout, with resilience being a quality that helps someone to recognise when they are at risk.

Examples of the value of resilience in coping with organisational stress were provided in accounts of various types of events. In particular, under neo-liberal policies of deregulation, work that previously required a social work qualification at times becomes opened up to workers who do not have any accredited education or training (compare with Trappenburg and van Beek 2019). This presents a challenge if services change in their nature, or social workers experience their professional culture being undermined. One social work supervisor described how shifts in government policy had caused such a change in a non-government organisation, which had been 'distressing'. Then, having changed its culture very quickly, the organisation had to make many staff redundant when the agency was not successful in the next round of tendering for the contract, so that also was distressing. In such circumstances, she thought, resilience was the quality that enabled her to maintain her own professional commitment.

Resilience is widely seen as a quality that can and should be consciously developed. This may be achieved intentionally, through supervision, training and other formal processes. In addition, one social worker commented:

> I believe in something called vicarious resilience. So, it's understanding that through my work I can actually be motivated, inspired, refreshed [...] I think you can look after yourself and have the resilience, but still be engaged and really care for the people that you're working with [...] [there can be] vicarious traumatisation [but] I developed vicarious resilience.

In this sense, resilience is strengthened by the everyday experience of practice. The examples that illustrated this understanding included disaster responses and also 'with a client where they weep or tell the most terrible things to you', as this social worker expressed it. In such encounters, resilience provides the personal strength to offer a compassionate response.

REFLECTIVE QUESTIONS/EXERCISES

What are the differences between perseverance and resilience?

Do you agree that resilience is a quality that can grow as well as weaken over time? Why/why not?

How do you see resilience in other people? How do you see resilience within yourself?

Connecting hope, perseverance and resilience

Throughout the discussion in this chapter we have noted that the qualities of hope, perseverance and resilience contribute to and sustain each other. However, this is not a simple or linear relationship. The statements of social workers, in particular, raise the question of whether it is possible to identify one quality of character as foundational to the others, or to suggest that one or the other is more significant in the ethical life of social workers.

To some extent, we may consider that without some form of hope, perseverance is greatly challenged. 'Keeping going' in situations where there is no hope is perhaps too great a demand for some practitioners. It may even be that when hope cannot be found, this is the point when most people would feel that it is the point to stop. However, how this can be experienced is also impacted by the possibility of resilience. The self-aware resilient social workers with whom the authors discussed this describe how, for them, resilience is experienced as a separate quality of character, even though for some it connects explicitly with hope and perseverance. From this we may say that the connections between these three ethical virtues are complex, with differences between contexts and the other qualities that any particular person has.

Cultivating hope, perseverance and resilience

Developing and sustaining hope, perseverance and resilience can be conscious processes for some people but not for others. In particular, being hopeful is an aspect of character that can appear to be innate for some people. Certainly, whether we consider the wider examples that are presented in the literature or the statements made by social workers to the present authors, hope is a quality that some people bring to their work, whether in micro or macro practices. At the same time, hope can be eroded, as can perseverance and resilience, when setbacks and challenges do not enable someone to see the possibility of change.

One of the ways in which social workers may develop and sustain hope, perseverance and resilience is from their life experiences outside their professional work. Upbringing and wider commitments (such as religious or political beliefs) play a role in the origin and continuation of the capacity to have hope. We have also noted that these three qualities also can grow as a consequence of the negative life events with which social workers have to assist people, not only in spite of them. This suggests that there is an innate disposition that some practitioners bring to their role, which then is strengthened in response to setbacks and challenges. These virtues can be vicariously built up, as we noted above.

Professional encounters also provide a basis for the development and sustenance of hope, perseverance and resilience. There is an important role for colleagues, supervisors and managers to recognise that even when these are innate dispositions, they can be harmed as well as nurtured through practice experience. They can also be brought out in people through conscious attention, so that supervision and debriefing, as well as explicit training, can provide opportunities for social workers to address the ways in which hope, perseverance and resilience can be recognised and strengthened. In training, although these qualities can be thought of as 'technical' virtues, and so taught as skills (Flesakar and Larsen 2012), as ethical virtues they are also open to being addressed in non-didactic ways. An alternative is suggested by Crowder and Sears (2017) with evidence from a mindfulness-based intervention. Yet another possibility is found in a report by Kiley *et al.* (2018) on a broader-based programme that utilised guided imagery as a means to assist practitioners to address perseverance and resilience. These are organised programmes that bridge personal development with a therapeutic approach. When such programmes are not accessible, simply making the virtues of hope, perseverance and resilience explicit may open up the opportunity for critical reflection.

The role of 'moral exemplars' can also play a part in this, by purposively looking at or revealing how others have worked with similarly difficult situations and maintained their commitment, irrespective of outcomes. In this way, the experiences of others who are regarded as good social workers may form a resource both for encouragement and as role models for the ways in which others have drawn on these qualities to develop and sustain their own hope, perseverance and resilience.

Conclusions on hope, perseverance and resilience

Much of the literature that focuses on hope and resilience in social work addresses these qualities as something to be nurtured in service users (compare with Pellegrino and Thomasma 1993; Banks and Gallagher 2009; Beauchamp and Childress 2013). Here, in contrast, our discussion addresses the way in which they are part of the ethical virtues that are relevant to social work (although it is the case that sometimes service users can develop greater hope, resilience and perseverance through seeing these qualities in social workers). Without them, remaining committed to a profession that involves close engagement with the problems and issues for which others require assistance is challenging and may even be overwhelming at times.

Each of these three qualities of character provides a complementary dimension to the ethical commitment of social workers and their capacity to hold on to that commitment. Together, they represent an insight into the moral strength that social workers bring to their professional role. From this, they also contribute to other ethical virtues, including courage, compassion and care, as well as justice, to which we turn in the next chapter.

FINAL REFLECTIVE QUESTIONS/EXERCISES

List three key learnings from this chapter.

Socrates' notion of self-care includes 'caring for one's moral self' (as discussed above). How might we both care for and continually develop our 'moral self'?

How can we nurture hope, perseverance and resilience, in ourselves and in others?

6
JUSTICE

CHAPTER OBJECTIVES

- To define justice as both a principle and as a virtue
- To explore examples of how justice is enacted in different social work practice contexts
- To consider how justice may be cultivated in the social work profession

The principle of justice

Social work has long regarded the concept of justice as a cornerstone of its professional ethics. It is clearly identified in the international ethical principles for social work (IFSW/IASSW 2018) and as a consequence in many national codes of ethics. However, it is important to note that when addressed in this way, justice refers to an ethical principle, rather than a virtue. In other words, justice is regarded as a value towards which social work should be directed. It is both a goal of practice and a way of understanding how social work should be practised. As a goal the value of justice is a claim about the desired condition of society to which social work ought to contribute and, as a way of practising, justice also refers to the relationships between social workers and service users, whether they are individuals, families, groups, communities or the society as a whole (in the form of research or policy work). From this we can say that the principle of justice requires that the outcome of social work should actively challenge injustice in the wider society, not simply avoid being unjust in itself. Thus, as a principle, justice requires that the ends and means of social work ought to be integrated.

In contrast to other professions, social work and related fields – which may be understood as the 'social professions', including youth work, human services and community work – widely refer to *social* justice (for example, compare Banks 2012 with Beauchamp and Childress 2013). Banks explains the meaning of social justice in these terms:

> [This principle] is based on the idea of distributing resources in society according to need (as opposed to desert or merit), challenging existing power structures and oppressive institutions and actions.
>
> *(2012: 54)*

This very clearly explains the primacy given by social work to make fair distribution of the resources required to meet need paramount in their theories and practices. Social justice is different from the retributive or restorative views of justice that predominate in legal systems. It also differs from other ways of understanding justice in the distribution of social goods, because it does not consider whether people 'deserve' such goods or have 'earned' them in a moral sense, but focuses on the needs that people have in living a decent human life, understood in their social context, or 'human flourishing' as it is called in moral philosophy (Papouli 2018: 4).

This explanation introduces another definitional problem, namely the question of what constitutes a decent or flourishing human life. Although this has long been debated by social workers, from its earliest days social work has tended to answer this question by focusing on negative and positive freedoms, that is, by considering what it is that people are free from and what are they free to do. An example of being 'free from' can be seen in the goal of assisting people to live lives that are free from poverty; an example of being 'free to' might be seen in the freedom to achieve the maximum potential that a person has, such as when a person with an intellectual disability is enabled to live independently, perhaps by providing the right support. In addressing these freedoms, social work focuses on the ways in which social inequalities prevent people from achieving them. This can be found both in people's own capacities to deal with challenges and in the ways social systems and structures create inequalities in access to the resources that people need to overcome such challenges.

In this distributional sense of justice, we can see the question that lies at the heart of Utilitarian ethics. This is an ethical approach that seeks the greatest 'utility' (that is, the satisfaction of needs) across the greatest possible number of people (Hinman 2013: 227). Utilitarianism has some value in helping us to consider the way in which the finite nature of the world always has to be balanced with an infinite variation in the demands that can be placed on it. This is the case whether we are considering material needs (food and water, shelter, safety) or needs that can be seen as psychological and social (health, education, culture). However, as MacIntyre notes, Utilitarianism has been widely criticised for leaving open the possibility that members of social minorities, or even specific individuals, might be harmed if this was seen to promote the wellbeing of the greatest number (MacIntyre 1989: 231–232). Some ways of distributing social goods are wrong for reasons other than 'utility', for example, when they require one person to give up something that is necessary for a good human life (such as their health) for a general increase in 'the common good' (Hinman 2013: 227).

One way of resolving this challenge has been to define such social goods in terms of 'rights', so that the claims of all those who lack access to the corresponding resources are legitimated (Wronka 1998). However, this shifts the point at which fairness becomes a concern; it does not remove it. When there are only certain quantities of such goods relative to the given number of people who need them, deciding between claims becomes a practical problem. In this context, social justice becomes a matter of ensuring that access to the achievement of rights is distributed fairly. However, the finite nature of social goods means that a way of deciding between claims is necessary, even though such claims are all equally morally valid. So it may mean that some rights are less strong than others.

In his analysis, Wronka follows Rawls, who begins his argument for a model of justice by noting that:

> Each person possesses an inviolability that even the welfare of society as a whole cannot override. For this reason justice denies that the loss of freedom for some is made right by a greater good shared by others. It does not allow that the sacrifices imposed by a few are outweighed by the larger sum of advantages enjoyed by many. Therefore in a just society the liberties of equal citizenship are taken as settled; the rights secured by justice are not subject to political bargaining or to the calculus of social interests.
>
> *(Rawls 1971: 3–4)*

In summary, Rawls' model argues that social inequalities can be defended only on certain limited grounds. These are (1) that inequalities relate to positions that are open to all on the basis of equal opportunity and (2) that in some way they are to the benefit of the least well-off (Rawls 1971: 5–6). Moreover, Rawls proposed that the basis on which inequalities in distribution might be decided should be seen as free from the personal interest of those making the decisions. Unless these requirements are met, under Rawls' approach inequalities are unjust. Rawls' theory has been criticised mostly on this last element, which assumes that impartiality can be achieved (Moller Okin 1989; Young 1990). These criticisms argue that it is not possible for those who exercise power and authority to put aside their own world-views, which are affected by gender, 'race' and culture, socio-economic class, age, (dis)ability, sexuality, religion and other social factors.

Nevertheless, social work has often implicitly followed the approach to justice proposed by Rawls, while also addressing criticisms about diversity (Clifford and Burke 2009). For example, a person with a physical disability who cannot attend a university because the facilities prevent them from entering the buildings is facing an injustice. So the challenge for professionals assisting that person is not a matter of how we might help them to change their mobility, but to get the building changed so that they can enter. In this instance injustice does not arise from the fact of the person's physical capacities, but from the social arrangements (including building design) that distribute social goods (in this case education). So under this approach social workers should be concerned with helping people who face disabling social systems by supporting them to change those systems.

On this basis, the principle of social justice commits social workers to challenge ways in which social systems operate. They may distribute life chances unfairly on the basis of factors that in themselves are not ethical grounds for such a distribution (such as gender, socio-economic class, race and ethnicity, sexuality, disability, age, religion, culture and so on) (Dominelli 2002; Clifford and Burke 2009). Thus, the foundation of injustice is seen to be social systems and structures, within which individual unjust decisions and actions arise. Moreover, social workers, and those in the other social professions, consider that achieving changes in the social systems and structures that underlie injustice is part of their professional role, even when they are working with individuals and families. They broadly agree that social justice is crucial to the ethics of social work.

O'Brien (2011) suggests that social workers may not have so much lost their way in pursuing social justice (as Solas 2008 argues) but that in neo-liberal conditions they find it hard to make social justice explicit as a core value that they are pursuing. Instead, by looking at accounts that social workers in Aotearoa New Zealand provided about their practice, O'Brien (2011: 177, 186) concludes that social justice remains a central goal but that it is being sought in social conditions that are hostile to its achievement. It is also the case that among the social workers with whom O'Brien spoke the majority were involved in services for individuals and families, including supervision and management as well as working directly with service users. This meant that their pursuit of social justice was often seen in practices such as advocacy, rather than in the pursuit of systemic and structural change (O'Brien 2011: 185). Although some of the social workers who responded to O'Brien felt that their opportunities to help people achieve socially just outcomes were limited, most continue to see social justice as a key value in their practice and to regard it as central to the ethical identity of social work.

So far, we have considered the principle of (social) justice. However, as identified in Chapters 1 and 2, our larger concern is with the ethical qualities or character of social workers. We need now to address the question of justice as an ethical virtue instead of as an ethical principle.

REFLECTIVE QUESTIONS/EXERCISES

How would you define a 'decent' life? What do you think is required for a human to 'flourish'?

How do you define social justice?

Do you think social justice has a place in social work, given the increasing challenges of neo-liberal approaches to social services?

Justice as a virtue

The earliest discussions of justice as a virtue occur in the writings of Plato and Aristotle in Greece (MacIntyre 1989). In their conception, the notion of justice referred both to the moral character of a person and the political environment of the state. In their conceptualisation, being a just person (virtue) in an unjust society (structure) was not possible, nor was having a just society without just citizens. For these philosophers justice meant the right balance within a society that included an acceptance of quite clear and at times gross inequalities based in gender, socio-economic class, 'race' and so on (Hinman 2013: 221). From the early-twenty-first-century perspective, this was understood as 'social harmony' in the sense of society being orderly rather than equitable. There are some similarities here with a Confucian view, although Lai (2006: 114–116) argues that it is a mistake to see Confucius' ideas simply as a form of virtue ethics. Although character is part of the Confucian framework, it is not directly comparable; nor is justice in an Aristotelian sense part of Confucius' concerns.

By the time of Aquinas writing in the European medieval period, two significant developments can be seen. First, justice as a facet of social relationships and justice as a personal moral quality had become divided (Pellegrino and Thomasma 1993: 93). Second, Aquinas derived his understanding from the interaction of classical Greek thought with the Judeo-Christian ethos that prevailed in his time. For this reason, he focused particularly on justice as a virtue that is held and expressed by each person as a reflection of their moral character.

For Aquinas, justice is one of the 'cardinal virtues', together with moral wisdom ('prudence'), balance ('temperance') and perseverance ('fortitude') (Pellegrino and Thomasma 1993: 4, 8). Cardinal virtues are those that are central to a person's moral character and enable the person to express their other virtues well. In this regard, for example, the qualities of social work may be said to reflect wider values, pointing to the moral qualities of social workers being an extension of rather than a divergence from wider moral standards. At the same time, breaking with the classical Greek tradition, Aquinas also held that a person could be just, and seek justice, even in a social context that was not just. We do not have to confine social work ethics within prevailing norms when these are asserted against the core values of the profession. Critical and radical perspectives within social work are centred on the possibility of practice that stands against prevailing social values when these are oppressive, exploitative and prevent people from achieving a good human life.

Nevertheless, relating the moral character of individual practitioners to the prevailing ethos of the surrounding society continues to be a concern in everyday life. Banks and Gallagher suggest that it is one of the more complex ethical concepts (2009: 161). Both Banks and Gallagher (2009) and Donaldson and Mayer (2014) refer to the way in which the ethical characteristic of justice tends to be ascribed both to individuals and to social systems and structures. Following Rawls (1971: 3), this leads them to refer to justice as an 'institutional virtue' (Banks and Gallagher 2009: 161) or a 'social virtue' (Donaldson and Mayer 2014: 208). This understanding is based on the idea that the foundations of justice are found in the qualities of the institution or society generally.

However, such a usage departs from the way in which the concept of ethical virtue is widely used in ethics (see our discussion in Chapters 1 and 2). This is unhelpful, at least for the present discussion. Justice may be seen as a 'characteristic' or 'quality' of an institution or the wider society, so that it forms an ethical aspect of social systems or structures, but these do not have 'character' in the way in which this is addressed by virtue ethics: systems and structures do not feel, think or act; it is people who do so. These are properties of individuals rather than institutions and social systems. (Insofar as systems and structures appear to have these properties, it is because they reflect the interests and choices of the people who create and act within them.) Banks and Gallagher themselves recognise this, noting that in virtue ethics moral character addresses 'dispositions, motives, emotions [and] moral luck' (2009: 39). This understanding leads them to define the virtue of justice as:

> a disposition to act fairly in particular situations (to calculate fairly how to allocate social goods and harms, based on sound criteria) and to promote and maintain fair social arrangements within the broad boundaries of the present social system.
>
> *(Banks and Gallagher 2009: 169)*

So, while they distinguish between social justice as the institutional quality of fair systems and procedures, and justice as a virtue that is an ethical characteristic of individual people, Banks and Gallagher argue that the two are connected. They conclude that a central quality in the virtue of justice is the disposition to seek, promote and defend social justice at the structural level.

Banks and Gallagher's explanation also rephrases justice as fairness. This is a complex idea because it requires a balance to be found between competing claims to social resources, in situations where not everyone is guaranteed to get all that they wish. Indeed, in those circumstances where it is possible to meet all needs, the virtue of fairness might not be required. However, as we have noted above, this is not the usual context of social work practice; social goods are inevitably in short supply relative to possible demand. Even in countries that have high levels of average household wealth, there are disparities that lead to some members of the society not having access to sufficient resources to have the capacity to live a decent (or flourishing) human life. One example in Australia is seen in the number of people who do not have sufficient income or access to housing at the level that is officially defined as the 'poverty line', and the disproportionate representation of families with children and of Indigenous Australians in such circumstances (ACOSS 2018: 23). In this context, social workers are often facing an ethical challenge when they have to prioritise who they will help and the goals towards which such assistance will be aimed.

Papouli (2018: 6) argues that instead of being only one of the cardinal virtues (of which the most important is wisdom according to both Aristotle and Aquinas), justice is the most significant virtue because it addresses the wellbeing or flourishing of others and not of ourselves. Justice is 'other-directed'. In macro-practice this quality might be seen in a concern for fairness in professional and organisational policies and procedures, or more widely of laws that promote wellbeing equally for everyone in a society. It may also be found in the practices of individual social workers who are concerned to be fair, insofar as they have the capacity to make choices in their actions. The point is that justice is the virtue that necessarily focuses our attention outwards and prioritises others in our ethical frame of reference. In the instance of responding to inequities in access to resources faced by particular groups of people, a just social worker would be one who is responsive to unfairness in the way that access to social goods is distributed in the society. It is this expression of justice as a virtue that O'Brien (2011) notes in his study, in the quality of fairness expressed when social workers talk about being frustrated with the obstacles they experience in being able to assist service users to achieve fairer outcomes.

Although much of the attention to justice in social work concerns social justice, Banks and Gallagher (2009: 160, 168–170) also observe that some social workers are involved in the criminal justice system, while some may be engaged in community work or in social development practice. In the criminal justice system, the just social worker may focus on whether the service user is treated fairly, as well as whether they have committed an offence and how they can be encouraged and supported to avoid future offending. Social workers have also sought to bring restorative justice into criminal justice, in which instead of retribution (punishment) those who have harmed others can 'make amends' in some way and so community relationships can be restored (van Wormer 2003). This practice is often called 'reparation' because it seeks to repair damaged relationships. Attributed to a Mäori practice in healing community divisions, this has now been adopted in other countries and cultures (van Wormer 2003: 444). The equivalent in social development can be seen in 'truth and reconciliation' processes that originally come from South Africa (Sacco and Hoffmann 2004; Androff 2010). In these ways of seeking to restore and support just community relationships, social workers express a just disposition. Where they have an active role, social workers also show justice in helping to facilitate reconciliation through the admission of harm that has been caused, and even in the possibility of achieving forgiveness that might come from this (Androff 2010: 1972).

Papouli (2018: 6) provides other suggestions of the ways in which just social workers can express this virtue. For example, how a social work lecturer treats students or how a practitioner treats service users in day-to-day interactions is a reflection of justice, especially given the inequalities of power and authority involved in these relationships. Similarly, for the social work teacher, being disposed to consider carefully how to recognise and respond to structural inequalities in the educational process expresses justice, for example, regarding gender in the classroom, or 'race' and ethnicity in marking assignments, or disability in obtaining accessible practice placements (practicum).

What these discussions of the virtue of justice in social work have in common is that they do not separate a person's disposition or qualities of character from their actions. Yet, in relation to justice, we might argue that the social worker who pursues social justice in helping service users to obtain particular resources, but who does not treat colleagues with dignity and respect (especially when they are junior in the sense of having a lesser status), is not demonstrating justice as a quality of character. More than this, although they appear to be pursuing social justice as a principle, how they are doing so is limited and actually raises questions about their understanding of the principle. Equally, if a social worker is just in their personal relationships with others as an individual but has no concern for social justice, we might ask whether that person is expressing the virtue of justice or some other quality. At the least, we might again conclude that their understanding of justice is limited.

There is a third possibility: that a social worker neither expresses justice in their own relationships with others nor is disposed to pursue social justice. This poses questions about how that person understands the ethics and values of the profession as a whole. As Papouli observes (2018: 6), justice is a significant virtue for social work because it is other-directed, and this includes the way in which social workers focus their practice on challenging injustice as well as how they act towards individuals. As a core function of social work is to attend to social needs, especially of the most disadvantaged, a social worker who lacks justice cannot fulfil their role ethically and, we contend, is unlikely to be able to do so in a functional (or technical) sense either.

REFLECTIVE QUESTIONS/EXERCISES

How do you 'show' or express justice in your practice? Think about your relationships with clients, colleagues and other professionals.

Does thinking about justice and its expressions in everyday life change your understanding of it? If yes, in what ways? If not, how did you come to your current understanding of justice?

The virtue of justice in practice

We have already noted that O'Brien (2011) concluded that although social justice does not appear in social work in the ways that it is often discussed in theoretical texts, it is to be found in day-to-day practice, for example, in the form of advocacy or in work that seeks to change service systems. The social workers with whom the present authors talked also frequently referred to 'justice' or 'social justice', both as the objective of their work and as a personal quality that they bring to practice. For example, one social worker said:

> I guess there's a few characters inside of me, and one of them is that riot girl that just wants to scream and shout [...] about whatever the injustice is in that particular minute.

Others mentioned their sense of anger or frustration at the way in which social systems and structures create inequities in access to social resources that people need to live a human life. Criticisms were made of the wider society, and especially neo-liberal public policy. As another practitioner expressed it, this is based on an:

> economic model [that] doesn't believe in recovery, they think if something's broken it's redundant, it's on the tip and we'll get another one [...] we have a consumer attitude to human beings.

For this person, social work itself is regarded as expendable within a neo-liberal framework precisely because the profession challenges this view of people. While this view addresses the principle of social justice, the social worker identified something about himself that drew him to social work and inclined him to want to integrate his own practice with this critique, not only in its goals but also in the way that he engages with service users. He connects his disposition to be just with challenging social injustices.

Similarly, when explaining social justice as a core value for the goals of her practice, another social worker observed:

> I relate to social justice, but you have to have a model for what justice is. We talk about democracy and fairness and equal rights and all that sort of stuff. There is inequality in the world [...]. It's mostly inequitable and unequal. Why do we bother talking about it if it's not there?

Here the definition of social justice gives a form to the sense that the practitioner has about her motivation to be a social worker, in her concern with injustice. Another social worker simply explained a sense of justice as 'being fair and just and equitable' in her relationships with service users and colleagues. Again, the disposition to be concerned about justice was regarded by a different practitioner as coming from an awareness of social inequities and feeling that 'this is wrong'.

> Social justice has always been what's motivated me to be in social work, and social justice is not really something that's talked about or dealt with I think in any way. Like, I would work very much around the social model of health, but it's not really the social model of health that has primacy in [the Department of] Health; it's more about a medical model of health, making sure their symptoms are stabilised and then getting them back out into the community as quickly as possible.

The suggestion in this statement that social justice is missing from practice is somewhat different to O'Brien's (2011) conclusion, although it echoes the challenges in social work that he notes. Among the social workers with whom the present authors spoke, there is a similar ambivalence – between the values of social work, the constraints of social and political contexts and the core dispositions of these social workers. Another practitioner commented:

> I've been reflecting lately about where exactly is social justice in our practice now, like, what's going on with that, and maybe – how do we, kind of, get that back a bit, that, kind of, 'stamp your footedness.' Because I think that social work more broadly has lost that a bit, that we don't necessarily – there's so many other things taking our focus that we're not as [concerned] about it any more.

For this person, justice was a key aspect of her own character, which had influenced her choice of social work as a career, and she could not understand what she saw as her colleagues' downplaying of social justice in their practice. However, she also recognised that this might be about how people express the quality of justice, which might be reflected in her notion of 'stamping your foot' as the right way to argue for social justice. However, while agreeing with the orientation to just practice, another social worker explained a less activist approach as a consequence of the limitations within current systems:

> For me, social justice is about fighting systems in a way, for want of a better way of putting it [against] things that are quite outrageous or unacceptable, or you know, tied back to [...] Liberal policy coming down that's just really going to be hugely problematic for client groups that I don't think we – there's so much of what happens in terms of government policy, for example, that we're so caught up in this scramble of work that we do that we don't then have the time to write submissions or to do some advocacy or protest in the way that I think [...] social work historically would have done.

However, another social worker gave a different account of why she takes a different approach, despite sharing the concern with injustice.

> [campaigning is] not my style, not something that I could ever see myself doing. But certainly advocating in smaller ways and being able to articulate a position on different things and to look for ways to shift around and be creative when perhaps there's something that I don't agree with. And sometimes biding your time until the time's right to be able to address it. I think somebody said to me 'you've got to pick your battles.'

Varying accounts are provided by social workers to explain their disposition towards just practice. One strong influence comes from the impact of upbringing, continuing family encouragement or the influences they have in the surrounding community and wider society. In this sense being just can be a cultivated virtue in the sense first discussed by Aristotle (MacIntyre 1989). It is a virtue because of what it inclines people to value, such as fairness in access to human rights, dignity and respect for everyone, challenging barriers and structural discrimination, which for many practitioners widely forms their original motivation for choosing social work. Some social workers also find their inspiration in engagement with influence from institutions or activities that explicitly hold a view about social justice. One example (echoed by others) is that:

> [religious faith] provides the foundational frameworks in which I operate the foundational understandings of social justice, of treating people the way that you want to be treated, of connecting with people who have been marginalised.

For others again, their personal quality of justice is sustained by a politics of practice. For example:

> social workers are best able to really understand how the socio-political environment impacts on our client and – and I think – so that person in environment, yeah, there's a whole range of theories in terms of [...] systems theories and all of those sort of stuff, but actually, interpreting that so that you can understand for a client what that means, that's the critical bit.

One of the injustices that Australian social workers face comes from the history of colonisation and its impact on Aboriginal and Torres Strait Islander people. Not only do Indigenous communities encounter poverty and oppression in much higher proportions than non-Indigenous Australians, but also there are specific events of the past that affect social work. The eugenic belief of the 1800s and early 1900s that Aboriginal and Torres Strait cultures were destined to die out produced a policy of forced removal of children from Indigenous mothers. These children have come to be known as the 'Stolen Generations' that continued into the 1960s, and together with other systemic exclusion and disadvantage this history has led to many Indigenous Australians facing high levels of social exclusion, disadvantage and oppression (Bennett 2019).

In their practice some social workers seek to respond justly to the challenges facing the profession. One practitioner talked about the dispossession that is encountered by Aboriginal service users. Although there were aspects of the situation that related to the way that non-Indigenous service users who have experienced trauma might have been assisted, this social worker also was angry at the overwhelming issues that came from a service user's Aboriginal identity:

> [...] there was a lot of really difficult issues going on, one of which was around him obtaining some housing and around proof of Aboriginality, and there was a lot of stuff within his [...] it was all stuff around access. And for me, there was something that happened within that case that really [I thought] 'this is all about dispossession and Stolen Generation,' there's all [this] intergeneration trauma stuff.

Another practitioner linked the disadvantage and oppression of Indigenous Australians with current systemic issues.

> [...] but I struggle with the systems abuse that I just see. And I think we do a great injustice to our Aboriginal people because I think that systems abuse is worse. And we lower their own expectations of what might be possible. And I do get extremely frustrated with people who don't believe that change is possible. Because I think why would we be here if we didn't believe that.

So for this social worker, the virtue of justice that is expressed in a rejection of the ethos of current systems is allied to other virtues, such as hope, perseverance and resilience. The disadvantage and oppression that challenge just social work include the following:

> Indigenous clients that need to go on dialysis and yet there's no dialysis chairs in their remote communities so they have to relocate to [the regional town, hundreds of kilometres away]. And families live in such poverty that they don't even have a fridge. Now that would be an outrage if that was in the city and that was a non-Indigenous family.

Throughout these observations, we see social workers expressing moral outrage (McAuliffe *et al.* 2010) that is an expression of their disposition towards justice.

At the same time as being motivated by the personal quality of justice, some social workers are conscious of the possible contradiction, given their own role in parts of the system. One practitioner in corrections noted:

> So I suppose really as a social worker you're dealing with the injustices of society but you're also an agent of the state so you need to do certain things like ring child protection or ring the police. So that can be challenging, balancing the power you have as an agent of social control at times but also trying to inhabit a position of working alongside families and being an advocate and it can be very uncomfortable.

In this critical reflection we can see that concerns with justice are not only about what others do, but also for the just practitioner must be applied to oneself. This concern can be seen also in the statement made by another social worker who provides groupwork for men who use violence in their domestic relationships, about the reality of professional power and the challenge this presents to the sense of self as a just person.

> [...] the orthodox training in social work where you're, you know, client-centred, you walk alongside them, you empathise, you acknowledge and encourage, and I guess the tension between that and also challenging them and pointing out that what they're doing is wrong I guess is not really in

> social work vernacular, that kind of ability too. Because then that's when you edge into the territory of [...] social control and of oppressed people and you're telling people what to do. So I guess navigating that tension, being able to [do so] with the client but also being able to challenge them in a way that's respectful and productive I guess that's not further oppressing them.

In this account, as in the one that precedes it, we see the way in which the disposition to be just is expressed reflexively in how the practitioner engages with service users. There are layers of injustice in such a situation, of gender as well as of 'race' and culture and socio-economic class, and the willingness to recognise that and respond to it shows a quality of character that expresses justice.

Considering all these accounts provided by social workers, it appears that there is a strong thread of justice expressed as an ethical quality of character. There are differing views as to whether social work is less concerned with justice than it once was, although our qualitative summary of statements from those with whom we spoke matches O'Brien's (2011) more quantified conclusion that social justice remains part of the ethical core of social work, even though it may not appear in the same form as it may do in radical and critical social work texts. Yet a personal sense of justice was explicitly stated by 42 of the 65 people (64.6 per cent) with whom we spoke as an important ethical dimension of who they are as social workers and the nature of the profession. It appears to be a morally significant reason why many have chosen social work and why they remain in what they report as a challenging field, one that is made harder by the injustice that they report as exacerbated by public policy and the lack of understanding of the lives of people who experience disadvantage and oppression.

REFLECTIVE QUESTIONS/EXERCISES

Did the pursuit of justice influence your choice to pursue social work? If yes, how did you develop a sense of justice as important? If not, what influenced you to pursue social work?

Reflect on the role of social work in pursuing justice for the Indigenous peoples of your respective country. What do Indigenous peoples say they experience as just?

Cultivating the virtue of justice

As we have noted, the quality of justice is one that is often brought to social work, having been developed through upbringing and sustained in family and community relationships. It is also strengthened through formal professional education, especially qualifying university programmes, although that is more likely to be through a focus on social justice that engages the moral self of the beginning social worker.

This suggests that there is scope for qualifying programmes also to address the sense of justice that students bring to their studies and to encourage a connection with their motivation for becoming social workers (Pawar *et al.* 2019). For those who come to social work with other motivations, addressing social inequalities, disadvantage and oppression can help students to discover this aspect of the profession's values and to integrate the principle of social justice, not only with their own view of what is just socially, but also with their reflexive sense of themselves as practitioners. Although they may be critical of the current neo-liberal ethos, social workers have long faced the challenge of making decisions about the use of scarce resources, even if that is how to prioritise their own time when there are competing demands that have equally compelling claims.

Banks and Gallagher (2009: 172) make two suggestions for professional education. The first is to use modelling of advocacy and the second is to use role-play of the allocation of scarce resources. Both these exercises, they suggest, can help students to get a clearer understanding of how studies in sociology, social policy, law and ethics will affect how they actually do social work, and so consider what it means to be just in their practice. Banks and Gallagher (2009: 171) also point to the importance of supervision and continuing professional education for beginning practitioners in the early stages of their career. This also points to the importance of the experience of the wider professional community in shaping and sustaining moral character. If, as we discussed in Chapter 3, good people can lose their virtuous qualities in certain circumstances, then avoiding the character of justice being eroded is made easier by the experience of working in a context where it is at least recognised and affirmed, even if it faces regular confrontations from the issues with which social workers assist people and from the prevailing ethos of public policy.

One way in which the experienced practitioner can maintain or even enhance their sense of justice is through involvement in collective action. This might be in a professional association or, where they are distinct, a trade union. Participation in a community organisation that campaigns with and on behalf of particular groups or issues is also relevant, such as with refugees and asylum seekers, mental health, the families of people in prison and so on. In these areas not only may social workers find that their sense of justice is supported and affirmed, or even heightened, but also that they can share their knowledge and skills when these are relevant.

Conclusions on the virtue of justice

Reports of the demise of justice as a virtue of social workers appears to be premature. Both research on the attitudes and beliefs of social workers about their practice and the evidence of social inequality that informs social work education point to the way in which concerns with social justice as a principle and justice as a quality of practitioners remain an ethical concern. Increasing neo-liberalism in policy and in political and public life certainly have increased the strains that social workers feel in maintaining their commitment to justice and to the pursuit of social justice goals.

At the same time, because it is a quality shared by many practitioners, social workers continue to look for ways to seek social justice and to achieve fairer outcomes for those who are disadvantaged and oppressed. Quite what this means in detail in particular situations remains highly contested, such as in the use of authority and control in situations where a person's actions compromise the needs and interest of someone else who is even more vulnerable (for example, in child protection). Nevertheless, even in these situations just practice is possible when social workers consciously address issues of justice for each person who is involved. In turn, this is possible more when social workers are themselves disposed to be just in their practice.

Even though current views of justice continue to follow Aquinas' separation of the just person and the just society, so that the former does not require the latter, there is a dynamic connection between the two. Having just social workers (and, indeed, just practitioners in other professions also) provides one way in which injustices may be challenged. Expecting unjust practitioners to challenge social injustices seems implausible; the more the society is unjust, the more unlikely this would be. To the extent that social work holds justice as a core value, if practitioners are unable to recognise and address injustice then the profession would indeed have lost its way, as its critics suggest (compare Solas 2008 with O'Brien 2011). However, the quality of a social worker being just is widely agreed to be a central virtue. It largely takes the form of fairness, which can support a range of practices that span from the equitable allocation of scarce resources at the micro level to arguments for a redistribution of resources more broadly at the macro level, so that more people in need are assisted. In this sense, the just social worker can exist prior to the just society (we might even argue that just social workers always have existed in this way; for example, see Payne 2005), but that envisaging a just profession without just practitioners, as with imagining a just society without just members, makes no sense.

FINAL REFLECTIVE QUESTIONS/EXERCISES

List three key learnings from this chapter.

How do you pursue justice in the face of challenges?

How can you nurture and continue to develop the virtue of justice, both within yourself and across the profession of social work?

7

HUMILITY

CHAPTER OBJECTIVES

- To define the virtue of humility
- To explore different views of humility in social work practice across different times and social and cultural locations
- To discuss the concepts of respect, dignity, altruism and expertise in relation to humility
- To present examples of humility in various forms of social work practice
- To consider how social workers can cultivate the quality of humility

The quiet virtue of humility

Understood as a virtue, humility appears largely to have been introduced into moral philosophy by Aquinas, in the European medieval period. MacIntyre (1985) argues that this concept was alien to Aristotle's thought. Indeed, he points to the way in which for Aristotle the virtuous person is one who is 'great souled', a man who is aware of his own many virtues and his status as a significant moral character in his community (MacIntyre 1985: 75). The gender-specific definition here is relevant, as for Aristotle virtues are qualities of character possessed by men of high social status. Women are excluded from this understanding, as are men of lower status. Indeed, Aristotle accepts the status of slavery and argues that some people are slaves by their inherent nature. By contrast, the 'great souled man' stands above others: 'He thinks nothing great. He only gives offense intentionally. He is very nearly an English gentleman. This appalling picture of the crown of the virtuous life [...]' (MacIntyre 1985: 76). For such a person, the concern of ethics is not in achieving a quality such as humility, but avoiding the vice of 'hubris', which is over-claiming one's greatness. However, in this way of thinking the 'great souled man' also corrupts his greatness if he fails to declare and assert it; humility is synonymous with false modesty.

In short, MacIntyre is viewing this early definition of the virtues from the position of a specific time and culture. His understanding of humility as a virtue comes from a different time and place, through Aquinas' arguments from medieval European Christianity (MacIntyre 1985: 114). Starting with the very different Judaeo-Christian notion that all people are morally equal, Aquinas adds humility to the Aristotelian typology of virtues, in which it is understood as the appropriate positioning of one's self morally in relation to others. From this, we can say that humility is about the recognition of the ethical claims of others, giving them equal moral weight to one's own and in corollary knowing that one is morally no better than anyone else. Thus, the vices relative to humility are arrogance and self-abasement (compare with Papouli 2018).

Historically there are few moral philosophers who pay attention to humility. Writing in the eighteenth century the Scottish philosopher Hume compares the moral emotion of humility with that of pride, but in doing so reframes it within his own emphasis on the role of emotion in ethics (Oakley 1993: 18). Hume uses the term 'sentiment', which in the early twenty-first century might be better understood as 'conscience'. However, Hume attaches meanings to these ideas that are the opposite of what Aquinas intended. By 'pride', he is referring to the positive regard we might have for something we did well or an achievement that is good. Conversely, when he writes about 'humility' Hume is referring to a negative sentiment in response to an act or achievement that we are aware is not well done or is even wrong – this is a response that today we might understand as 'shame'.

The most significant departure from Aquinas occurs in the twentieth century with Nietzsche's concept of the *Übermensch* (often translated as 'superman' or 'overman'), who is the person who asserts their own moral code and shows their superiority through the exercise of power over others (MacIntyre 1985: 113–114; Robinson 1999). Nietzsche sees Judaeo-Christian ethics as a 'slave morality' because it values being other-directed and holds humility to be a virtue. He arrives at this point because he rejects both the claim by Aquinas that morality is objective because it derives from a common human nature and Hume's opposing argument about shared moral sentiment. This leaves Nietzsche only with the possibility that morality is what people make of it for themselves through an heroic assertion of their will (which he calls the 'will to power') (Robinson 1999). MacIntyre, conscious of the use of these ideas in the horrors of the Holocaust in twentieth-century Europe, calls the idea of the *Übermensch* an 'absurd and dangerous fantasy' (1985: 113), although he also notes that Nietzsche himself also attacked racism and was not nationalistic, anti-Semitic or a political fascist. However, the *Übermensch* bears a curious likeness to Aristotle's 'great souled man', a likeness that does not sit well in the increasingly democratising modernist era and is easily interpreted in racist and nationalistic ways.

As we have already identified in Chapter 6, social work is other-directed in its ethical goals as well as its social function. In this context humility resonates both intellectually and emotionally, understood as the appropriate moral positioning of one's self in relation to others and so granting equal moral weight to all ethical claims. Social work has long held clear practice principles such as 'starting where the client is', 'self-determination', '(moral) non-judgementalism' and 'empowerment', each of which in some way concerns placing the service user on an equal moral level with the practitioner (Trevithick 2012). Yet, perhaps somewhat surprisingly, there has been relatively little attention paid to the virtue of humility in social work ethics until quite recently. So we now need to turn to the specific way in which this virtue has been addressed in the profession. The following discussion will also show how humility relates to other values, including those that may also be regarded as virtues, such as respectfulness, dignity and altruism.

REFLECTIVE QUESTIONS/EXERCISES

What are the 'vices' related to humility?

How might we balance self-belief (and/or confidence in our professional skill and knowledge) and humility in social work practice?

Humility in social work

In reviewing the attention that has been paid to the ethical virtue of humility in social work, we can identify three distinct trends. First is the acknowledgement of humility without fully developing this as a distinct notion, which we can see as rendering it as an underlying or background virtue. Second, there is a clearly identifiable way in which humility is related to practice, policy and institutions in what might be termed colonised or 'settler' societies, such as Aotearoa New Zealand, Australia, Canada and the United States of America. Related to this approach, the idea of 'cultural humility' has also emerged as a more general expression of this virtue in multicultural societies. Third, there is a smaller discussion of humility that seeks to develop a more generalised understanding, either directly or through associated qualities such as respectfulness, dignity or altruism. We will consider each of these in turn.

Humility as an underlying virtue

Discussions of humility as a background or underlying virtue are evident in discussions of ethical dilemmas and the contradictions between social work's (principle-based) values and agency policies and procedures. For example, Weinberg (2009: 149) notes that without humility social workers cannot involve service users in participatory decision-making. This, she argues, requires that social workers recognise that many ethical questions do not have a single, easy answer but are often more or less good. Answers to such questions may even combine good and harm, which she identifies as 'moral trespass' (Weinberg 2009: 147). Elsewhere Weinberg notes that social work ethics requires 'the solidarity of a community that recognize[s] the inescapability of trespass, allowing for humility, doubt and clemency' (2010: 41). By this she means that in order to be open to service users wanting to take risks that risk-averse systems try to prevent requires that social workers accept that their knowledge and skills have limits and that contradictions cannot, as such, be resolved but rather present stark choices. To do this means that social workers must accept their doubts and permit themselves 'forbearance', or compassion towards one's self (Weinberg 2009: 149).

One such doubt is identified by Marston and McDonald (2012), who argue that the tendency in social work scholarship and teaching to emphasise the pursuit of social justice through social change as the core of social work leads to an 'heroic struggle' view of the profession, especially on the part of students and recent graduates. In contrast, they propose a 'more humble' approach to the sense of political agency that holds onto social justice as the goal but accepts that steps towards it are messy, complex and often unfinished (2012: 1034–1036).

Similarly, Houston (2011) locates the importance of humility in the dilemmas and contradictions of complex practice situations. Houston (2012: 660) argues for a more explicitly pluralist ethical understanding, in which values are frequently 'incommensurable' and 'incompatible' (that is, cannot be compared and cannot be made to fit together). Pluralism here refers to the ethical approach that sees value in a number of approaches, rather than taking a position that there is one correct (principle-based) approach. This latter position is called 'monism'. The challenge of pluralism for social workers is that choices must be made to act without the certainty that there is one right action, compared to the appeal of monism in principle-based ethics (whether deontological or consequential), which gives the appearance that one best way of acting can be found. For Houston (2012: 667), accepting that this is the case requires ethical humility, as practitioners grapple with the tensions between intellect and emotion, and between key principles such as freedom and equality. Like Weinberg, Houston wants this to be a creative tension in which there is room for self-doubt and self-criticism.

Beddoe *et al.* (2019) report qualitative research with social workers in Aotearoa New Zealand in which practitioners were invited to discuss their perceptions of themselves as social workers. The research was centred on a question of whether social workers have pride in their occupation, in the Humean sense of pride as the positive feeling of satisfaction that one's actions and achievements are good and right. Although there was a strong positive response, in that practitioners spoke of 'making a difference', 'contributing to wellbeing' and 'being an agent of change', there was also a frequent expression of humility offered as an alternative to pride, contesting pride as sometimes being an expression of arrogance and boasting (Beddoe *et al.* 2019: 538–539). Beddoe and her colleagues note the way that some social workers talked about the opportunity to share their skills and knowledge with people who are disadvantaged or marginalised as a privilege. They also quote one social worker who said social work 'is a humble profession; it's not meant to be about us' (Beddoe *et al.* 2019: 539).

Between these three arguments we see a commonality in the way in which the quality of humility enables social workers to practise well, not only in the sense of achieving professional values such as empowerment and participation, but also in placing a concern with service users at the centre of their ethical engagement with their work. In addition, humility here is regarded as the acceptance that expertise is not fixed but reflects the fluid, ambiguous nature of the social problems and issues with which social workers struggle alongside service users.

REFLECTIVE QUESTIONS/EXERCISES

Define ethical pluralism and monism. What are the differences between these two approaches to social work ethics?

Think of a time in your practice or training when you reached the limit of your professional skills/knowledge. How did you recognise you had reached this point? How did you respond?

Humility in cultural context

Social workers writing from, or about, practice with Indigenous peoples refer to humility as a virtue in practitioners who are not themselves Indigenous (at least to the area in which they are practising). For example, from a Native American background, Weaver (1999, 2017) refers to humility as a central quality in social workers who are willing to learn from (not about) Indigenous service users. Without doing this, Weaver argues (2017: 143), social workers will continue to practise in ways that perpetuate colonisation and fail adequately to pursue the values of social justice or human rights. In contrast, humility opens up the possibility of listening and responding to Indigenous peoples. Weaver also makes the point that the virtue of humility thus becomes a basis for good social work with all peoples.

Voss *et al.* (1999) make a similar point when they observe that expressions of humility are deeply culturally embedded. Non-Indigenous social workers seeking to work with Indigenous service users face the challenge of understanding that the cultural roots of humility are not only intellectual and linguistic but also emotional, psychological, spiritual and moral. By taking the culturally dominant (that is, global Northern) ways of being and understanding as universal, social workers are unable to engage appropriately with Indigenous people and so at the least simply fail to connect, and at worst perpetuate what can be regarded as 'clinical colonialism' (Voss *et al.* 1999: 233). Voss and his colleagues add the point that some actions that are considered to be expressions of virtues in European cultures can be precisely the opposite in Indigenous cultures. For example, in a European cultural context, engaging eye contact when speaking with another person shows respect because it conveys that they are the focus of attention. However, this is seen in entirely the opposite way in Native American cultures, as in that context age, gender and other status characteristics may often make maintaining eye contact an act of arrogance and disrespect (Voss *et al.* 1999: 231). This same difference is often found between European-derived cultures and African and Asian cultures as well as Indigenous peoples, and there are other similar differences such as volume and tone in speech.

Walker, a Mäori social worker in Aotearoa New Zealand, goes further when he says, '[w]e have a role as social workers, educators and researchers to be "not knowers", that is humble experts who serve others. Remembering at all times that "the learning is in the relationship"' (2012: 70). Walker grounds this understanding in a critique of 'cultural competence', in which social workers can believe that working with Indigenous peoples or cross-cultural practice can be achieved by knowing 'about' other cultures (Walker 2012: 69; compare with Dean 2001). The same problems are encountered in Australia for Aboriginal and Torres Strait Islander people in their experiences of social work and the wider human services system (Green and Baldry 2008). Russ-Smith (2019: 110–112) explains the relationship of humility and respect in Wiradyuri culture (the Indigenous people of what is now known as south-west New South Wales) in the ethics of 'yindyamarra', meaning to live with deep respect in a world worth living in, through nurturing right relationships with the country, the community and other people.

These Indigenous ethical frameworks are not simply abstract ideas that can be learned intellectually. As with other ethical virtues, 'knowledge' in this sense is when a different way of understanding and acting in the world becomes an expression of a person's being. The historical and structural context is also very important. Without a self-critical awareness of these power relations, cross-cultural social work risks becoming a form of racism, however unintentionally (Dominelli 2002). Humility in this sense involves being open to allowing yourself to be changed by the encounter with others. In summary, by listening to Indigenous peoples and reflecting on cross-cultural practice, we can understand humility as the quality of confidence in the relationship between what one knows (professional knowledge) and what one does not know (service users' knowledge of their own culture and life), in which the former is placed in a balance with the latter.

A comparable perspective can also be seen in critical thinking about social work between people of different national cultures. For example, in international social work uncritical universal views about human nature are questioned as forms of professional imperialism or neo-colonialism (Midgley 1981; Gray and Fook 2004; Gray 2005). In this field of social work we can also see calls for cultural humility, through recognition that dominant (global Northern) professional ideas and values are derived from the cultures within which they developed. This is a call for 'cultural humility', through acknowledgement of the way in which assertion of encultured values as 'universal' perpetuates colonial relationships. For example, the African values of *ubuntu* and *ma'at* both embody the idea that in order to live well, people should understand themselves in balanced moral relationships with the physical world, the community and other people (Graham 2002). In this way, by giving respect and dignity to others and acting with the needs of others in mind, the people acting also achieve respect and dignity in themselves.

> **REFLECTIVE QUESTIONS/EXERCISES**
>
> Think of your own cultural and social context. How are humility and respect routinely conveyed?
>
> What are some of the limitations with the 'cultural competence' approach to working cross-culturally?
>
> How can humility be utilised in approaches to cross-cultural social work?

Humility in relation to respect, dignity and altruism

These considerations of humility point also to the way in which this virtue is connected to other ethical qualities. In particular, respectfulness and the promotion of dignity are sustained by humility, again because they are other-directed. Altruism, the virtue of being willing freely to give or do something for another person without any expectation of receiving something else in return, is another ethical virtue that links with humility.

Respectfulness and the promotion of dignity are addressed by social workers not only in the way in which work with service users is conducted, but also in policies and in the structures and systems of service provision. Banks (2012: 42–47) points out that the notion of respect as it has been understood in social work primarily draws on Kantian ethics, which is principle-based, universalistic and formal. In that framework it refers to valuing the equal moral identity of all human beings and of acting towards others on that basis. This accords with the virtue of humility and is expressed as an ethical virtue in respectfulness towards other people. For example, in qualitative research with service users, Chadiha *et al.* (2017) identify several elements in respect, including esteem (for role or status), affirmation and deference, expressed in different ways such as through providing care and support or seeking advice. From a literature review, they also note that others have included acceptance of difference and recognition of moral equality.

Hölscher and Bozalek (2012) explain respect in terms of offering other people dignity in the way they are treated. They note that a lack of respect and the failure to provide dignity often arise from misrecognition (not accurately perceiving the identity or needs of the other person), maldistribution (or resources to meet need) and negative moral and political preconceptions of the other person (2012: 1108). Looking at these questions from the perspective of social work with refugees and other vulnerable groups, they argue that respectfulness and dignity are possible if social workers and other professionals are able to grasp their own privilege and find ways to relate with those they seek to assist (2012: 1107). Likewise, Steen *et al.*

(2016) connect respect and dignity as ethical qualities that are expressed by being provided for service users. While Hölscher and Bozalek, writing in South Africa, further link respect and dignity to the value of social justice, Steen and her colleagues in the USA see them as characteristics supporting the pursuit of human rights. In both instances dignity is regarded as the quality of being seen and treated as a human being, with the same moral worth as all other human beings.

As an ethical virtue on the part of the social worker, respectfulness enables dignity to be shown to others, with implications of equality of moral worth irrespective of status and identity. To achieve this, the ethical virtue of humility as we have defined it here provides the grounding, because it concerns social workers placing themselves in precisely that relationship of equality of moral worth with those around them. Although this seems to present a problem for social work because of the many areas in which authority and power are necessarily exercised, we agree with Banks (2012: 47–48) that this is an inadequate grasp of ethics. For example, the notion of 'non-judgementalism' means not writing the service off as morally unworthy of assistance; it does not mean resiling from challenging people about actions that society deems to be unacceptable, but rather engaging with the person as a human being, with strengths and capable of achieving change. Or, likewise, respecting confidentiality provides a person with dignity, even though their actions may have led to social constraints such as a court order mandating intervention.

Altruism is a complex ethical virtue and one that some social workers consider unsuitable in a professional context. Indeed, emphasis is often placed on the maintenance of appropriate boundaries, of guarding against risks such as dual relationships or of burnout and so on (Reamer 2018). More widely, however, altruism is highly regarded. The classic example is found in those countries where the donation of human blood without payment is the normal way in which health services maintain a supply of blood (Titmuss 1971). The criteria for an action to be altruistic are that it should be done without compulsion, that it should be without reward and that the lack of it will not result in a person being regarded negatively in ethical terms; again, this is a quality of being other-directed.

Altruism is rarely discussed in social work, although it is sometimes mentioned briefly in other professional literature, such as health ethics (Beauchamp and Childress 2013). Beauchamp and Childress mention the choice of career as a possible aspect of altruism, when practitioners purposefully choose less prestigious areas of work when they could have done otherwise (2013: 51–54). Some social workers certainly do make such choices. For example, there are fields of practice that can be more or less favoured, such as work with children and families being preferred over care for older people (Duffy and Healy 2011). However, we suggest it is a consequence of the nature of most social work practice that these are not necessarily perceived as altruistic; rather, such choices are seen as normal for the profession and as expressions of social workers' interests and aptitudes. Thus the sense of 'moral excellence' in altruism that Beauchamp and Childress ascribe to medical practitioners who choose less prestigious fields would mostly not be understood by social workers in quite the same way.

In the different ways that we have discussed here, the ethical virtue of humility disposes social workers to relate to service users in particular ways and so to seek to achieve other values, including social justice, human rights, respect and dignity. We have emphasised that humility is not a matter of someone denying their own moral identity, or although it may involve 'putting other people first' at times. It concerns the development of recognition and valuing of the moral identity of others and achieving an equitable balance between arrogance and self-erasure.

REFLECTIVE QUESTIONS/EXERCISES

What is the relationship between respectfulness and humility?

Consider the criteria for an action to be classified as altruistic. Do you think altruism has a place in professional social work? Why/why not?

Humility in practice

For the social workers with whom the present authors discussed ethical virtues, humility was expressed in several ways. The two main ways it is encountered are in seeing every person as morally equal and expressing this in actions towards others, and in being open and honest about yourself and the social work role. Social workers also understand humility through expressions of respect and the promotion of dignity of others, especially service users.

Humility in the form of seeing everyone as morally equal is conveyed in this short statement by a practitioner:

> I think humility is important, a humility that you're just as human as the person you're serving, you're not better than them, in fact it's your privilege to be helping this person.

This echoes the finding of Beddoe and her colleagues (2019) that is discussed above, in which humility does not simply relate to how a person sees her or himself, but also the positive regard that is given to others that values the other person. In this, we can see a moral foundation for other-directedness.

Other social workers expanded on this, elaborating the way in which these different characteristics are connected. In the context of addressing a service user, this quality is expressed in the following terms.

> I would say that I'm no better than you as a person, even though I have a university degree and I have a university qualification. I live a middle-class life, you have no university qualification, you have little high school, and you're struggling on a low income, [but] I don't consider myself a better person than you. Sitting down with you I'm learning a great deal from you

> about human society, about human relationships, and that's rewarding for me. And you know, I want to be thankful to you for helping me understand more about life.

In this way, humility is seen as a foundational aspect to ethically good relationships with those whom social workers are serving.

The way in which humility takes the form of being open and honest about yourself and the social work role takes several forms. First, it can be understood as demystifying what is happening for service users, explaining processes, showing service users notes that are taken in conversations, consciously learning from others and in the styles of communication that are used. This is graphically illustrated in the example provided by a practitioner who described her work in this way.

> I was trying to, I was going to say 'be humble' [...] I'm not sure if that's a good word. But I suppose some of the things that I did was to engage [...] they're basic but I would sit on the floor of the woman's room, I wouldn't sit in a chair above her, I'd sit on the floor. I would use language that I think was easy for her to [understand]. I was transparent. And I also had this hope that people can tell if you're genuine and I think really, particularly in certain cultures people can pick up on non-verbal ways of being. So I was also just hoping that she might pick up that I was, had compassion for her and respect and I was hoping that even little things like making her a cup of tea would be, that she'd experience that as supportive.

In this context, humility also weaves together other ethical qualities such as compassion and care. Attention to another person's cultural ways of communicating and trying to convey a sense of actively listening are also important elements.

A supervisor extends these ways of expressing equality, openness and other-directedness when she says:

> Somebody actually said to me at one stage that one of the things they like about my supervision sessions was my humility. And I said to this person, 'What do you mean?' because we often use [that] terminology and word, 'What do you mean by my humility?' And then he said, 'Look, it's your capacity to be humble about the things that you have done well but also to be open about the things you haven't done well.' I don't think that I had actually realised until that time what I had been doing. Oftentimes it takes somebody else to tell you.

This suggests that humility is both more easily recognised in another person and also that it is considered more easily when another person draws attention to it. Social workers do not readily tend to claim this ethical virtue – which is perhaps consistent with the virtue itself.

Humility is also clearly connected with the characteristic of respectfulness. This can be seen in practice in a number of different ways. It can include paying attention to other people, listening carefully to what people say, and being able to do so in a way that the other person can recognise. Respect for others is also shown in both verbal and non-verbal communication. As one social worker expresses it, humility combines with respectfulness through:

> listen[ing] intently to a client, without interrupting the client. Which I know from [...] personal experience [...] a lot of people find that difficult, actually. And so clients have often reflected [...] that they've never been listened to quite in that way [...] people don't often have that opportunity of being in a room, just two people, it's quiet, and it's just you looking at the client and the client's looking at you. Like it is a unique setting [...] and I actually even set up my room to be not too distracting and having it quiet and having it comfortable.

In a different context, for another practitioner the same quality is expressed in making clear that control of the conversation is shared between the social worker and the service user. The emphasis is on the importance of the service user in this encounter.

> I find it a really honourable place of respect when I can say to somebody, 'I don't need to know it all,' and I think that they really appreciate that, and I actually, I think those are those little moments that I feel proud of, where I get to do that for somebody, whereas, to protect that dignity, or protect that information, I don't need to know it, and sometimes I'd rather not.

So, in this way the humility of seeking equality in the communication enables respectfulness to be demonstrated.

In the same way, the ethical characteristics of humility and respectfulness combine with empathy and compassion in other-directedness. One social worker considered that:

> in my practice in the way that I show respect to people, in the way that I value what's significant to them. You know, I think sometimes it's intuitive about not pushing a particular point.

A different social worker expressed the same idea in describing the characteristics of a respectful interview in:

> maintaining that person-centred focus. That I let him be in the driver's seat for our dialogue. I didn't try and force myself or my beliefs – professional beliefs – on him, but I allow him to dictate the conversation, or allow him to dictate what he wants to do, when he wants to access certain services, how he would like my support or not.

Again, taking a broader view of practice more generally, another person commented:

> So I see it as a dialogue we're all learning from each other. Which is what recovery-based community work is [...]. It's not about putting things on people, it's about dialoguing with people in community and learning together and travelling together.

In such encounters, the service user is valued and is able to experience this, because the social worker is communicating respect through humility in their approach to the relationship.

From these expressions of ethical humility, social workers also identify moral non-judgementalism as a quality that can be held and communicated. For one practitioner, this also enabled a strengths-based approach to be meaningful.

> So I think it's easy as a worker just to get caught up in how people are viewed, so I think it's really important to hold true that they're people who've been impacted by some unfortunate circumstances, and that it's not helpful to be classifying them or labelling them. Yeah, we're really seeing them as an individual [...]. And I think just being non-judgemental is massive. But to be actively non-judgemental. And, just curious in how, what a survivor she was, and I think finding the strengths in her story was quite different to how she'd been viewed. [...] Well I think it's easy to say we're non-judgemental if we're not saying anything but sometimes what we're displaying can be non-verbal but it can still be judgemental. So I think if you're not being judgemental you actually have to be the opposite of it.

The important element in this encounter is that the social worker is herself aware of the way in which non-verbal as well as verbal communication can express moral judgement. Similarly, for another social worker humility simply enables her to see the service user as a human being.

> That stuff around dignity, worth, self-determination – that people can make mistakes and not be punished as naughty clients. You know, that we value people's lived experience and that we take that seriously and it's not just about being a nice warm fuzzy social worker, this is about how people should just be respecting each other as human beings.

Beyond an individual interview or conversation, the non-verbal dimensions of being other-directed are evident in various ways. A person who keeps their promises, for example, demonstrates that they are valuing the other person. The practitioner who seeks actively not to do any harm as a minimum ethical objective also is displaying this quality. One of the aspects of practice that social workers identified that illustrates this is in working with people of different cultural origins to their own, especially when the service user is from a marginalised and disadvantaged background and the social worker is not. For example, for one person this is summarised simply as:

> [r]ecognising my own privilege, I've had a fairly privileged white middle-class upbringing, and so recognising that [...] a good social worker treats people with respect, empathy and compassion but also has really strong social justice principles that certain things are not acceptable and continually fights [for what is right] [...] A good social worker breaks down barriers that people have in accessing to do what they want to do and a good social worker is culturally safe across all cultures but also gender, disability, everything.

Also acknowledging the importance of such recognition, another practitioner expands on this point in saying that in work with an Aboriginal man she had recognised in herself the disposition to listen to and value the strengths gained from that person's culture.

> [D]espite Australia's brutal practices of colonisation and trying to strip his dignity, that was inspiring to see how he would talk about using his cultural knowledge to support him in the recovery from his stroke. So getting connected back to the land. And that despite, I suppose, colonising practices which try and convince people that their knowledge is less than or not adequate or not the norm, that he was still very connected to his cultural knowledge and values.

Here we can see that humility and respectfulness can be qualities of the way in which someone engages with cultural as well as individual 'otherness'. In this way, these characteristics can be seen to be relevant in macro as well as micro practice. A social worker in services for asylum seekers gave a comparable account of the moral response that is evoked in him, when he described a particular asylum seeker who:

> just held onto such a strong dignity and yet the system was, the way it was responding to him it was just pulling that away, it was just destroying any sense of dignity that he had. And I would sit in meetings with Immigration and the Health Service just saying he needs to be out of here [i.e. advocating that the asylum seeker needed to be out of the detention centre] and they're going 'there's no medical reason for him to be transferred' [...]. He was an amazing person in himself and his family and so I learnt so much from him in terms of how to hold onto dignity and your sense of self despite everything that's happening around you.

Again, a different practitioner echoed this way of seeing the relationship between their own moral disposition towards service users simply as fellow human beings and the ethical issues that follow from that.

> Everyone has a right to be treated with dignity and respect and safety, and not to be in a violent situation. And I guess another layer of that, and again as me as a social worker, seeing that my role is micro and macro, so working with the client to help them reconnect with their passion for their rights [...] but also macro in looking at ways to systemically change the world.

For most of the social workers with whom we discussed these perceptions there is little sense that they attribute their sense of humility, or the related characteristics of respectfulness, to a particular basis. There are those who consider a spiritual or faith orientation as part of their ethical identity in this regard, which implicitly makes connections with the Judaeo-Christian origins of the inclusion of humility among moral virtues by Aquinas that we discussed early in this chapter. For example, one social worker spoke quite generally of the moral equality of the relationship between herself and a service user.

> [Y]ou know, I just find it really a sacred place [...] it might be to do with God, although I'm not really too convinced that it is, just, I think it's just that recognition of how special people are, and that some moments are just really special and need to be recognised as such.

Another person saw this in terms of his background and the way that his overall ethical self has developed. He noted that he grew up in a particular religious framework that shaped his core values, including the sense of the moral equality of all people:

> within that religious community and within my family that are consistent with social work and they're things of respect and dignity, and that we need to carry those as an individual person and an individual practitioner. I'm not associated with that religion now, but those values very much became a strong part of who I am. And I think social work has been informed by – to a certain degree by Christian religious values.

This value base also formed the grounds for a clear criticism of those who do not bring humility to social work, as he added:

> but I don't necessarily see a lot of social workers adhere strongly to some of those fundamental values – respect and dignity. I think I've seen a lot of social workers who are more interested in their own [...] pursuing their own career up the hierarchy of the organisation, and they'll step on anybody to get there [...] I've worked with, for example, many senior staff who think they're better people because they're further up the hierarchy, that they're smarter, they're more intelligent – whereas to my mind they're not, they just happen to be further up the hierarchy.

This illustrates the point that humility does not of itself prevent someone from asserting what they hold to be ethically right – indeed, to allow it to do so would be to risk the vice (in virtue ethics terms) of self-effacement. Again, having identified humility as a characteristic of a good social worker, another practitioner commented:

> that's an expression of where my Christian spirituality is at the moment. That I want to see workplaces where there is a high level of diverse orientation and background and us all learning from each other and respecting each other and me learning from others as well [. . .] that reflects a change in my own theology and my own ethic of social work as well. Yes, so my own religious beliefs wouldn't be advertised here unless people ask me.

Here, too, someone can be confident of their own value base while accepting and valuing the differences experienced by others. In this way, each of these illustrative examples reveals ways in which someone can be aware of the source of their own ethical stance while at the same time being able to respect the differences between their own and others' beliefs and ideals, where these are consistent with a broader understanding of social work ethics and values.

REFLECTIVE QUESTIONS/EXERCISES

How do the qualities of honesty and humility relate to one another?

One of the practitioners in the above examples discussed the notion of being 'actively non-judgemental'. What do you think this means, and how might this look in practice?

Think about your own practice/training. What are some significant learnings that you have gained from clients/the voices of service users? (Note: service user perspectives might be conveyed via research, consumer advocacy channels, media, written submissions and so on, not only via direct client contact.)

Cultivating the virtue of humility

The philosophical approaches to humility from Aquinas or Hume, the more recent scholarship on the appearance of humility and pride in social work and the illustrative comments of Australian social workers together point to an important aspect of this virtue: that it is developed through a person's life. Humility is expressed as part of a wider world-view, related to other values and virtues. However, while it may be formed through specific beliefs, such as spiritual or religious or political beliefs, it is not confined to those who explicitly express such views. This suggests that for many social workers, if they possess the ethical virtue of humility they bring it to their practice rather than learning it as part of the professional curriculum.

At the same time, it is also possible that as an ethical virtue, humility can be acquired through experience. This can occur through encounters with service users. There is a clear contemporary recognition that service users ought to be regarded as experts in the realities of their own lives (in strengths theory, the recovery model and in other person-centred concepts) and this both supports and values humility in practice (compare Saleebey 2002 with Bland *et al.* 2015). To this extent, humility and the associated qualities of respectfulness and promotion of dignity can be supported and encouraged in education and training, but they need to be grounded in other aspects of the ethical self of the social worker.

Humility is also modelled by colleagues, who in this sense may be regarded as 'moral exemplars' (Pawar *et al.* 2017). It may be that those who demonstrate the opposite qualities are successful in organisational terms, as illustrated above by reference to achieving seniority at the expense of others. However, the congruence of humility with other social work virtues such as respectfulness, or practice values such as moral non-judgementalism and promotion of self-determination, makes this a more powerful ethical quality as a model for those who aspire to good practice, where 'good' is understood in both ethical and technical terms, integrated in action. For all these reasons, we can say that ethical humility can be enhanced and supported by conscious influences through education, training and explicit discussion in the profession, but it arises from the more complex interaction of wider experiences and values that social workers bring to their practice.

Conclusions on the virtue of humility

In this chapter we have discussed the ethical virtue of humility. Although this is not a characteristic of professional life that is widely discussed, either in social work or in other comparable caring or human service professions, we have seen here that it is very closely related to core values of social work as well as to the ethical virtues of respectfulness and promotion of dignity. This appears to stem, at least partly, from a misunderstanding of humility as 'shame', 'self-effacement' or some other quality that from a virtue ethics perspective can be considered a vice (in that it is a deficiency of appropriate self-regard). Rather, the meaning of humility that we have argued for is in a balanced view of the moral equality of all people, on the basis of which social workers may regard the human dignity and moral worth of service users positively, without losing sight of their own value. The value of a social worker may still lie in the knowledge and skills she or he possesses, but they are offered to service users as opposed to being forced onto them, even in circumstances of exercising statutory authority. The social worker with humility listens carefully to the service user, acknowledges their views and concerns, and responds with respectfulness. In Sercombe's telling phrase (2010: 10), as professionals we 'do not provide a service, we serve'. Humility is the ethical quality that enables this to be a reality.

FINAL REFLECTIVE QUESTIONS/EXERCISES

List three key learnings from this chapter.

Do you agree that humility is an important characteristic for social workers that enables us to pursue 'good' practice? Why/why not?

How might humility be cultivated, nurtured and further developed in social workers?

8

PRACTICAL WISDOM

> **CHAPTER OBJECTIVES**
>
> - To define the virtue of practice wisdom
> - To explore the virtue of wisdom, and its applications to social work practice
> - To discuss the concepts of intuition, evidence, critical self-awareness and reflexive practice
> - To consider how social workers can cultivate practical wisdom

The virtue of wisdom

Social workers place great importance on 'practice wisdom' (Flaskas 2007; Hunter and Kendrick 2009; Clark 2011; Banks 2016). However, this characteristic has more than one meaning. It can refer to the way in which some practitioners develop the capacity to make connections between aspects of a situation so that they reach an understanding of the issues without the need to work laboriously through various stages of analysis. In this sense, wisdom is the distillation of a range of experience, including formal education and observation of practice, that can be applied to new situations in ways that are seen as appropriate by others who are involved (for example, Flaskas 2007: 195). Practice wisdom can also refer to the ability to engage in practical reasoning, in which the various steps of ethical argument are carefully examined and the best possible outcome is then pursued (for example, Banks 2016: 44).

Hunter and Kendrick (2009) helpfully observe that wisdom does not depend solely on specialised forms of knowledge or experience possessed by professional experts. For them:

> [t]o exercise wisdom one cannot solely consider knowledge but also the context of its use, the values it upholds or subverts and the role of larger, more super-ordinate truths that transcend empiricism.
>
> *(Hunter and Kendrick 2009: 160)*

From this, they continue by noting examples where the lived experience of service users as the basis for peer support among patients led to better health care outcomes than reliance solely on the highly technical knowledge and skills of professionals. It is not that the professional knowledge and skills were 'wrong', but that there was another dimension in the knowledge and skills of service users that enabled them to contribute what professionals could not.

The central question for our discussion is whether wisdom should be considered an *ethical* virtue, or be regarded as a technical quality. For the classical philosophers, the answer is complex. Aristotle's use of the concept is to regard practical wisdom (which in Greek is *phronesis*) as that quality in someone that enables them to make sound ethical judgements (Pellegrino and Thomasma 1993). Pellegrino and Thomasma explain this in the following terms, noting that for Aristotle *phronesis* is:

> the virtue of practical wisdom, the capacity for moral insight, the capacity, in a given set of circumstances, to discern what moral choice or course of action is most conducive to the good of the agent or the activity in which the agent is engaged. *Phronesis* is the intellectual virtue that disposes us habitually to attain truth for the sake of action, as opposed to truth for its own sake.
>
> *(1993: 84)*

In this sense, then, wisdom is the virtue that enables us to exercise good judgement in the use of all our ethical virtues. It is the quality of being aware of how our moral and technical virtues related to what is good and right in any given situation, and to choose actions that respond appropriately. Banks and Gallagher (2009: 83–84) note that wisdom consists of the way in which technical knowledge is brought together with 'professional artistry' in the quality of the capacity to understand the complexities of a given situation. It is for this reason that from an ethics perspective we are concerned with *practical* wisdom, and not simply wisdom for its own sake.

Again, as with other aspects of virtue ethics in professional life, our contemporary considerations of practical wisdom have reached us through other bodies of thought. Banks and Gallagher (2009: 77–78) observe that the major world religions all contain reference to the quality of wisdom as central to the objective of being a good person. In their summary, they also identify particularly the connections made by Judaism, Christianity, Islam, Buddhism and Confucianism between wisdom and action (also see: Peterson and Seligman 2004; Hunter and Kendrick 2009). In the Judaeo-Christian tradition in particular, following Aquinas, practice wisdom is understood as 'prudence', in which practical wisdom as the quality of recognising and ordering ethical virtues with action is allied with the disposition then to undertake those actions (Pellegrino and Thomasma 1993: 84–85). Practical wisdom must lead to practice. For Lai (2006: 120), the same is also true for Confucian ethics, in which wisdom is not what one has but what one does.

The term prudence has lost the significance that it may have once had, through increased association specifically with a particular religious perspective and, although reference is still sometimes made to prudence, in recent discussions of this virtue in the professional literature there has been a revived use of the term practical wisdom, and even at times *phronesis* (Banks and Gallagher 2009; Papouli 2018). In this discussion, therefore, we refer to practical wisdom throughout.

Also often carrying a sense of being 'outdated' is the ethical virtue of temperance. Especially from a global Northern perspective, this term is widely associated with social movements against the consumption of alcohol and other drugs and other actions that may be pursued for pleasure, but which are seen by critics as harmful to individuals and to society generally. However, for the classical philosophers, temperance is a cardinal virtue (MacIntyre 1989). In that understanding it refers to the disposition to achieve a balance in a person's life between excesses in the pursuit of what is pleasing and avoidance of what is seen as harmful or unpleasant (Morales-Sánchez and Cabello-Medina 2013: 729). Thus it may be regarded as the characteristic of moral balance. Again, for the modern world the medieval European Christian interpretations of Aquinas have had a significant impact on our grasp of this virtue (Pellegrino and Thomasma 1993: 119). There are also similar approaches in other world religions, including Judaism, Islam, Buddhism and Hinduism (Peterson and Seligman 2004; Hunter and Kendrick 2009; Hugman 2013).

Temperance also refers to the combination of other ethical virtues, including honesty, humility and care, where these dispositions are moderated in the ways in which they are expressed. It may also serve to moderate justice, in the sense of inclining a person to forgiveness, mercy and moral non-judgementalism. In this sense, like practical wisdom, temperance is seen by some as a cardinal virtue because it is the quality of being able to balance the other virtues as well as the pursuit of things that are desired (Morales-Sánchez & Cabello-Medina 2013: 728). For Pellegrino and Thomasma (1993: 120–121) temperance also refers to the way in which professionals can avoid the pursuit of power over others. It is not that professionals can avoid the exercise of power and authority; what Pellegrino and Thomasma are referring to is the possibility of domination, in which professionals do not enable service users to be as self-determining as possible, but instead seek to control others' lives on moral or political grounds. In this way, temperance comes close to being a form of practical wisdom, as it enables professionals to achieve balance in their approach to those they serve.

Hunter and Kendrick (2009) make the point that practical wisdom and temperance are not characteristics seen only among professionals, but (as with all the ethical virtues discussed in this book) can be regarded as part of what might be considered 'ordinary' or 'everyday' ethics. Nevertheless, even though we may agree with this point, in the context of professional life ethics also takes particular forms and has a distinctive significance (Banks 2004: 65–68). While the values, principles and virtues claimed by professions come from the same ethical ground as does what might be called 'the ethics of everyday life', in the context of professional work there are particular meanings that follow from the specific knowledge, skills and functions of professions. Our primary concern here is the way in which virtue ethics relates to social work, in this case practical wisdom and in relation to that temperance, so we now turn to a closer consideration of the recent debates in this particular context.

REFLECTIVE QUESTIONS/EXERCISES

Given the various ways of understanding the term as explored above, what is your preferred definition of practical wisdom?

Do you think practical wisdom is a virtue, a technical skill, or both?

Consider Banks and Gallagher's (2009: 83–84) consideration that wisdom is where technical knowledge is brought together with 'professional artistry' in order for a social worker to understand the complexities of a given situation. Do you consider social work to be a form of art? Why/why not?

Practical wisdom as a social work virtue

Practical wisdom is not only a concern for social work. Recent attention to this virtue occurs also in comparable professions such as nursing (Arries 2005; Newham 2015) and school teaching (Culham and Bai 2011; Cooke and Carr 2014). There are many similarities and agreements, not only in the understanding of the idea of practical wisdom and what it means in general terms, but also in some of the broader ways in which it supports practice. However, given the importance of context for the exercise of practical wisdom, in each instance the way in which it appears is somewhat different.

Papouli (2018: 6) presents a general argument for the importance of practical wisdom in social work practice. The examples that she gives include knowing your own limits and knowing the limits of rules and principles. In the case of the former, we may refer to the ethical concept that if someone undertakes actions for which they are not adequately trained this is, in itself, ethically (as well as potentially technically) bad practice. It is dishonest, denies the right of service users to informed consent and so potentially exposes them and others to unacceptable risks. Wise practitioners know their limits and are confident to ask for help from others who are qualified. It is in this sense that we can say competence is an ethical issue.

Awareness of the limits of rules and principles is somewhat different. This aspect of practical wisdom prepares the practitioner to struggle with ethical challenges, engaging in ethical questioning rather than always expecting codes and procedures to provide formulae that offer ready-made solutions to the messy and often indeterminate world of people's issues and problems. Many situations in which social workers are involved do not have a single, easily identified ethical solution.

Although there are many ethical aspects of any circumstance that can be readily identified (such as honesty), at the same time there are some that require wisdom in how they are expressed. For example, in truth telling the practitioner may need to take account of the impact of bad news or a challenge to a person's hopes, so *how* the truth is told becomes an ethical question.

There are also situations in which equally compelling values cannot be pursued equally, for example, when there are conflicting rights and needs between individuals, each of which will be affected by the actions of the social worker. In such a context there may need to be an ordering of values; moreover, that ordering will be different in each specific situation (Houston 2011; Banks 2012). The term 'ethical dilemma' is used widely in social work to describe all circumstances where there is not one immediately unambiguous ethical position that is 'correct'. However, here we are seeing a dilemma as being a situation in which this ordering or balance of values, principles and virtues has no single, clear resolution and the hard work of reaching a position that is acceptable has to be undertaken.

Banks (2016) presents an analysis of the complexity involved in wise decision-making. In particular, Banks associates practical wisdom with one aspect of the process she calls the 'ethics work' undertaken by the morally active practitioner (compare with Husband 1995). For Banks practical wisdom is seen in the pursuit not of reason, as such, but of 'reasonableness' (2016: 43). That is, the practitioner is not seeking to follow generalised or abstract universal rules, but to make decisions in ways that achieve the best balance between competing ethical aspects of the particular situation. The judgement of what is 'best' in any particular circumstance is related to who is involved as well as other surrounding factors. At the same time, Banks notes that practical wisdom involves more than being able to reach a reasonable position in any one situation, but also is found in bringing together all the dimensions of 'ethics work', which includes making sense of events and circumstances, grasping the role required of the practitioner in different situations and one's identity in a specific situation, engaging with emotional aspects of the situation, forming relationships with others and then undertaking the work appropriately (which Banks calls 'performance work'; 2016: 45). A key point is that although practical wisdom is perhaps most visible in the achievement of reasonable understanding and actions following from that, it also encompasses every element in 'ethics work' as Banks describes this (2016: 43).

It is in this way that practical wisdom supports and enables the other ethical virtues to be expressed in the accomplishment of social work (McBeath and Webb 2002). However, as McBeath and Webb point out, we are not looking for a generalised, unitary sense of what practical wisdom might look like. Banks' (2016) notion of 'ethics work' is not an instruction manual that is intended to be followed in a series of stages that will lead to a clear 'wise outcome'. Banks (2016: 37) actually rejects the idea of a prescriptive ethical framework that instructs a practitioner on exactly what steps to follow to produce the correct answer to an ethical challenge. Her concern is to assist social workers to develop a greater understanding of ethics, in the theories that inform the widely shared norms that define the profession and in the realities of making sense of these values, principles and virtues in complex practice situations. In this understanding, codes of ethics are the vehicle for prescriptive guidance, but even in that context it is still necessary for practitioners to particularise from the general statements of codes – they cannot be construed as detailed instruction manuals.

Although Banks (2016: 37) asserts that 'ethics work' is not the same as 'moral fluency' (Sellman 1996), which she describes as a skill in dealing with ethical concepts, she also acknowledges that 'ethics work' requires 'ethical fluency' because it enables the 'doing of ethics' that is 'ethics work' (also compare with Husband 1995; Hugman 2005a: 163–165). This basis of relationship, between the work of ethics (doing) and fluency in ideas (knowing), lies in the way social workers have the disposition to exercise judgement, reflecting who they are (their 'being') (Pawar and Anscombe 2015). In this sense, practical wisdom can be seen as bringing together doing with knowing and being, in what is sometimes referred to as 'praxis'.

In a way that is consistent with Banks' viewpoint, Adams (2009) argues that practical wisdom (as with the other virtues) is primarily developed through practice experience. He makes the very important point that even when a social worker applies a decision-making framework or asserts a code of ethics, it is still necessary for that person to exercise judgement about 'what inputs from the environment are relevant or salient' (Adams 2009: 86). The 'decision-making tree' approach to doing ethics does not take away the importance of being able to choose between the elements that are included and considered. Making such choices is the exercise of judgement, and that in turn can only be done well if it is done wisely. For this reason, we cannot dispense with the need for practitioners to cultivate practical wisdom through developing their ethical self in social work.

Looking at practical wisdom in this way indicates the importance of social workers reflecting critically on all aspects of their work and their own actions within it (Gardner 2014; Pawar and Anscombe 2015). Although Gardner does not address the term wisdom explicitly, she identifies the way in which making sense of ethical questions in practice requires continual reflection on a person's own moral self and the relationship of the self to decisions and actions. The point of critical reflection is that it not only concerns being aware of what ethical values and qualities guide someone's practice, but also helps in avoiding what she describes as 'unthinking moral certainty' (Gardner 2014: 130). It is not that a person's own values and qualities are to be abandoned, but rather the capacity to consider their own position enables practitioners to be able to give an account of why they hold a particular perspective or to change their position – in short, it enables them to be ethically accountable to themselves as well as to others. For Gardner, *critical* reflection requires a questioning of cultural and structural relationships and assumptions in the way ethical values are formed and held (2014: 22–23). This is especially important when someone's own perspective is at odds with the formal ethical statements of the profession (for example, in a code of ethics) or even more so is challenged by an employing organisation's policies and procedures.

For Pawar and Anscombe critical reflection, drawing on cultural and structural self-awareness, can be seen as achieving a 'higher level' of understanding of the professional self in context (2015: 45). Because of the messiness of practice and the particularities of any given situation, they also emphasise that it is the overarching quality of practical wisdom that enables practitioners to continually re-evaluate their responses to people and events as work proceeds. Understood in this way, practical wisdom draws on past experience that has been carefully reconsidered, not simply to apply this as a model for future work, but rather as the basis for knowing what further questions might be asked, as well as awareness of how someone's own thoughts, emotions, values and so on affect how they hear and see what is being said and done and then react appropriately.

We have already noted that the ethical virtue of temperance may be seen as closely connected to practical wisdom. If temperance is understood as the ethical virtue of self-control that achieves balance in a person's life, clearly it is supported and enhanced by critical reflection. Papouli (2018: 7) provides four examples of how temperance might appear in social work: working hard without compromising your health or personal life; creating appropriate boundaries to limit excessive demands; where possible keeping your personal life separate from work; and achieving a balance between healthy activity and pleasure. These all contribute to the notion of 'self-care' because they assist social workers to look after all aspects of their self, including social, emotional, relational, physical, psychological and spiritual aspects.

For Papouli (2018: 6) this virtue – moderation or self-control – is a means by which social workers might avoid illness, exhaustion, burnout or even making mistakes in their practice. We can observe that there are practical ways in which the exercise of moderation or self-control requires judgement; for example, exactly where should we draw the line between working hard to address issues and problems to which we have a strong ethical commitment and taking a break to care for our selves? Many issues and problems that call on a response from social workers are urgent for those whose lives they impact and may be very demanding. Making judgements about such boundaries or limits requires a form of practical wisdom, so that the sense of moderation is seen in not being inclined either to withdraw too soon or alternatively to risk sacrificing oneself. Papouli (2018: 7) refers to the idea of particular virtues being a 'hinge' (the linguistic root of 'cardinal' to describe key virtues) and in this sense practical wisdom is cardinal because it hinges with temperance to create both the disposition to seek balance and the capacity to achieve it. Likewise, temperance is a cardinal virtue because it forms the basis on which social workers might achieve a balance between the demands of the other virtues.

REFLECTIVE QUESTIONS/EXERCISES

How can the virtue of temperance inform how we set boundaries in our work?

When it comes to ethical decision-making, what are some of the limitations of codes of ethics?

Practical wisdom in practice

Intuition and evidence

In the conversations with social work practitioners that we are using to illustrate our discussion of ethical virtues, one of the recurring themes in relation to practical wisdom is the idea of 'practice wisdom', as distinct from evidence-based practice as a quality of mature professional work. Social workers even speak of 'practice-based evidence' as a counter concept to 'evidence-based practice' (compare with Flaskas 2007). Nevertheless, evidence-based practice has become a movement in professional work, especially in the health professions, even at times being referred to simply by the letters 'EBP' (Smith 2004).

It can be argued that there is a strong logic to support the emphasis on evidence-based practice. Put succinctly, why would a social worker practice in a particular way if there were no evidence that that particular action would have a better effect than doing nothing, or doing something else? For example, Macdonald and her colleagues (1992) argue that what often pass for 'wisdom' are untested assumptions or even ideologically driven conscious ignorance of systematically tested evidence. Against this is the concern that the range of factors ('variables' in research methodology) is too great for the degree of certainty this argument proposes, except in particular areas of micro-practice with individuals and nuclear families (Smith 2004; Gray *et al.* 2009). Although this latter position does not seek to dispense with systematically tested knowledge, it also advocates tempering this with the need for judgement based on practice experience to deal with the 'indeterminacy' of the everyday world – that is, practice wisdom that is able to discern the subtleties of service users' lives and their social contexts.

Examples of the statements by social workers that illustrate this point can be couched in terms of this debate. For instance, one practitioner said:

> not necessarily what the book tells us is right, but what our experience is of what works, and what is right, and sometimes that takes a little bit of like testing and getting it wrong, sometimes too [...] evidence has been defined as research evidence in the quantitative, gold standard, experimental versus control group approach, which works in some things, but it negates the life experience that you've had as a worker.

At first sight this may appear self-defensive, or even to deliberately ignore systematic evidence, as the value of systematic evidence is that it challenges 'everyday wisdom' that may be incorrect. However, this person went on to discuss the way in which research addresses tendencies and norms, which many people's lives do not fit. Applying knowledge to the differences of particular individual situations requires judgement. Another social worker amplifies this point in noting that there are also constraints on the way in which decision-making occurs, when she observed that social workers should not:

> presume that you have all the facts. Don't ever presume that you have all the facts, because when you speak to the family [I work in a medical setting] there's a whole world of things that haven't been asked, and [we are] very time-poor. But don't assume that you have the information that you need.

So, under these circumstances there is a need to exercise judgement based on a combination of appraisal of what is known and experience in what has worked previously.

Some social workers regard systematic evidence and theory as the foundation for professional judgement.

> I think what influences my practice most of all, and I recently went back to do a Master of Arts in social work, and I was studying in the area of relationship-based practice. So I think that very much informs my practice [...] that's where you set the foundation for your practice.

And again, from a different practitioner:

> it was wonderful to have [a student] through [named] university to get back to the theory underpinning, that this is not 'common sense.' There's actually theoretical underpinnings for everything that we do and to realign with that.

Both of these statements acknowledge the distinctive contribution of practice based on knowledge and ideas learned formally through education. At the same time, for others the basis of practice relies on a more intuitive use of accumulated experience, which is articulated as 'gut feeling'. For example:

> [So] for the first few interviews I just keep listening and, you get your gut feeling [...]. So maybe we can say I'll go an instinct-based approach or something like that, but definitely just listen and [...] make sure that rapport is built with a [conversation] and respect.

This conveys a sense of practical wisdom guiding how an assessment is achieved and the way that future work follows from that. Similarly:

> for me the intuition is it's not the verbal it's the environment, the mood, the body language, the other signals that I'm getting from the person that is, helping me to understand where they're at. So it's almost as though I'm connecting with them, on an intuitive level [...] again listening to the person and trying to be on their level with them.

Throughout these conversations, social workers referred to 'listening' as a central dimension to practice wisdom. The wise practitioner is one who listens carefully.

Social workers also combine the exercise of intuitive judgement, based on the accumulated experience of practice, and evidence and theory. One practitioner graphically expressed this point when she said:

> as a girl of the 1980s where we just said 'Oh my gut told me it was good so I'm going to do it,' I said we've got to move on from that. It's great to have a gut reaction but, and that's where in fact that woman who was our first supervisor or leader who'd been at the [names agency] who said 'Look I understand gut reaction,' she said 'But I actually want to understand' [...] [so when a colleague says] 'Oh my gut reaction was' I go, 'Yeah, yeah, that's fine. So what was it that was upsetting your gut?'

Another social worker described the value of evidence and theory as providing a map for the practitioner to follow with the service user. In this sense, intuition (the 'gut reaction') is a very important part of practical wisdom, while at the same time testing this against evidence and accumulated experience enables the practitioner to be critically reflective and to avoid being reactive or following any unintended bias.

REFLECTIVE QUESTIONS/EXERCISES

How do you utilise each of the following in your practice:

- intuition
- theory
- evidence?

How would you describe or define a 'wise' social worker?

Critically reflexive self-awareness

Another dimension to practice wisdom is the way in which social workers are aware of what they bring to their practice in terms of their own ideas, knowledge, skills, values, commitments and so on. This includes their own understanding of particular theories and methods of practice. For example:

> I am – and I know it's not that popular – eclectic. I do use a range of different theories; I do cobble things together in different ways. I suppose, I can see over the course of say ten years here that I'm using very different things than what I used ten years ago, but I think that's experience. And so I suppose I have more of an attachment and trauma focus in my work than I would've done ten years ago.

This statement points to the way in which the social worker is open to learning and developing their practice based on the evidence that has accumulated over recent years in relation to the way that issues of attachment and the experience of trauma affect the lives, relationships and actions of service users (Howe 2011; Levenson 2017).

From a somewhat different perspective, another social worker gives an account of critical reflexivity as practice wisdom.

> I think that having a passion or a commitment to a particular community or a particular demographic of people has assisted me [...] I also think trying to establish a collaborative relationship with families and ensuring that you're transparent in your practice and also really continuing to critique the operations of power that exist in your workplace and in your role, I think is really important. So thinking about [it], I suppose the professional privilege that you have as a social worker and being aware of that and being careful not to use it in an inappropriate way or just being conscious of it really.

The practice wisdom in this instance is expressed in *how* practice is undertaken, as compared to the previous example that concerns the evidence and theory being used to select appropriately *what* is done. The wise social worker is conscious of professional power and authority and how this is used (compare with Hugman 1991).

What we are suggesting here is that the virtue of practical wisdom that enables social workers to practise 'well' should be seen in the way that experience assists the practitioner to bring together the best skills, knowledge and values, through their critical self-awareness of who they are, what they seek to achieve and how they undertake their work. Thus Papouli's (2018) point about a cardinal virtue such as practical wisdom being a 'hinge' between other ethical and technical virtues is shown to operate.

REFLECTIVE QUESTIONS/EXERCISES

What is the relationship between practice wisdom and critical reflection?

How has your practice/learning developed over time? If so, in what ways have you adapted the way you practise as a social worker?

Practical wisdom and social structural challenges

Social work often requires practitioners to engage with people who are unlike themselves in various social structural dimensions of identity and experience. These include gender, ethnicity/culture and 'race', socio-economic class, age, sexuality, (dis)ability, religion and so on. Being able to engage with those who do not share your own social background and life experience requires not only knowledge and skill, but also the practical wisdom of being able to understand the other through empathy. Various examples are provided by social workers, such as in a description of an encounter with an Aboriginal Australian family by a non-Aboriginal social worker, in which the latter learned about a culturally relevant way of supporting the family.

> There was one lady, an Aboriginal mother pulled me up and said, 'You stand here,' and I just stood, I didn't do anything else I just stood there. And this mother, the grandmother then was yelling at her daughter and pointing at me and saying, 'That welfare lady, you know she's welfare so she's got to do [describes statutory intervention].' And then she goes, 'Now you can go' [...] all she wanted was that authority for her to berate her daughter to stop drinking to look after her children, 'I'm going to make sure that you're going to look after your child, because if you don't this fat white lady is going to take your baby.'

Somewhat differently, another social worker described the difficulty of reconciling issues of gender in cultural difference faced by a social work student she had supervised. The issue was in a request to provide domestic assistance for a family because the mother was unable to do certain tasks when there was a man also in the household who could do them. The family argued that he could not perform the tasks because in that culture men do not do these things. The social worker framed her challenge to the student in the following way.

> Do you think we should respect that culture and provide them with cleaning service because that is their culture and perhaps that's something we need to respect? Or [...] on the other side do you think the government should be cleaning for people who don't believe that men should clean? It took her ages [...] I suppose I present that as a dilemma, and I'll often present dilemmas and then, I said no the government is not going to provide.

Here practical wisdom is reflected in recognising that the question of the interplay between gender and culture is significant for the family, but at the same time policies prevail in this type of decision-making. That is, the values represented in the policies refer to disability and not gender as the basis for allocation of these resources (men as well as women can be expected to undertake domestic tasks). In that way the social work role becomes one of representing to the family the dominant value perspective as embodied in policy, so that they understand the basis of the decision. The dilemma lies in understanding that this is contentious because, for some people, it represents a form of institutional racism, whereas for others it demonstrates a problem of sexism.

For some social workers there are more general or diverse challenges across multiple structural distinctions. When commenting on the need to be respectful, one social worker added:

> being respectful of the basics of people's religious and cultural differences, their age differences. You know, I've got three clients who are aged 80 or more, right down to my youngest is probably early twenties, [they are] across the spectrum. But being respectful of people across the whole range of [characteristics].

In these circumstances, which describe social work in many places, practical wisdom lies in the capacity to be respectful while also working with the constraints of organisations and policies that often reproduce a particular view of the world. Practical wisdom can also take the form of knowing when to stick with the established processes, when to challenge them and when to use ways of moderating the power of institutions and professionals in order to respect everyone involved, while still being able to achieve outcomes such as the protection of the most vulnerable people in a given situation.

REFLECTIVE QUESTIONS/EXERCISES

Consider the actions of the social workers in the two case scenarios above (one that involves a non-Aboriginal social worker working with an Aboriginal family, and one that involves questions around gender roles, culture and the provision of domestic services to a family). How do you think you would act in such situations?

Think about a situation in your own practice or study where you were working with someone quite different from yourself (difference/s may be across gender, culture, 'race', class, (dis)ability, age, sexuality, religion, etc.) How did you convey respect and empathy? What were the challenges?

Temperance as a form of wisdom in practice

Social workers are also concerned about achieving a balance between the demands of their work and their own needs. One way in which this is expressed is in a physical sense of 'looking after one's self'.

> [L]ittle things, like having a cup of tea […] just something to enjoy or look forward to. I have a blanket in my desk, when I want to just feel cosy. There's a concept, and I've forgotten the name, because it's a Nordic word […] they have really warm hot chocolates and things are soft and the colours are chosen to be just so, just cosy, and it's this concept of really being so aware of your environment and the way you speak to each other, and the company you keep, and the atmosphere that you are putting around yourself, to really foster this kind of homely warmth, this deep-seated joy and comfort […] just fostering things around you that make you happy, and be completely selfish about making yourself happy, because someone has to.

This has parallels with the classical Greek idea that the 'the care of the self [is] required for right conduct' (Foucault 2000: 285).

A different approach to balance is seen in the way that other social workers set boundaries for themselves around various aspects of their work. This includes how relationships at work overlap with time not working. As one practitioner expressed it:

> your colleagues are your colleagues, they're not your friends. And really being clear about who you invite in your home because we live in the small communities, it's having really strong boundaries about work relationships and family relationships, because there's a lot of enmeshment that occurs with relationships in small communities.

For some people this may appear to be quite rigid, although the context of life in 'small communities' (as opposed to large cities) is clearly the context here, in which different aspects of life (work, family, community and so on) are not separate but usually blend into each other. As with the previous example, the question is how to set boundaries to care for the self, in this case socially, culturally and emotionally rather than physically.

Different again are those who seek balance in the way that they engage with the content of their practice. This involves cognitive and emotional boundaries.

> I don't own the problem myself. So that's how I kind of manage it, [I] used to churn and get really upset about stuff, but now [...] you can't protect everybody [...] you become desensitised and so my threshold is quite high. So it has to be really bad for me to be upset.

Similarly, another social worker observed:

> I think self-care is a really key one. I think just remembering that people have the right to make their own choices and that [...] my key role as a social worker is to help them to make the decision that's best for them, not necessarily that I perceive is best for them [...] and that that's very, very tough sometimes.

Again, a different practitioner commented:

> a lot of times [...] social workers care very, very deeply and they see a lot of the trauma of the world, but they absorb it [...] that's highly dangerous, because what it does is creates this loop of irresponsibility, because the practitioner isn't taking responsibility for their own health and wellbeing, and looking after themselves, and therefore they're not able to support the client to take care of themselves [...] [then] we wonder why burnout is so extreme.

So, through these different observations we can see that the achievement of balance, the virtue of temperance, involves practical wisdom in the way in which social workers seek to care for and protect themselves against the ways in which a lack of balance (being drawn into others' problems and issues, never being away from work) both harm the self and also impact negatively on their capacity to assist service users. Care of self is balanced with care of others.

REFLECTIVE QUESTIONS/EXERCISES

What are your thoughts about the boundaries between work life and personal/home life? Do you feel there should be a clear separation between the two, or that boundaries should be permeable? Why?

How do you care for yourself in relation to your work? Think about self-care in terms of physical, emotional, cognitive, ethical, practical and spiritual.

Cultivating practical wisdom

The qualities and capacities that comprise practical wisdom require development, which in turn needs to be gained from experience. Although some social workers talk about 'intuition' as the basis for their ability to find their way through the indeterminacy of everyday practice, this is not an uncritical or untested 'common sense' but rather it is a thoughtful and considered reflection on what has worked and if and how the same approach might work again in a different situation. Insofar as practical wisdom arises from professional maturity, it does not simply happen as a result of time spent in practice, but it takes conscious attention and critical reflection. Effective supervision and learning from colleagues in day-to-day situations are part of this process, but it is also always the case that social workers must themselves be engaged with critical reflection, because what each social worker contributes to any part of practice is also key to whether previous learning is used well, or whether it simply becomes an uncritical habit. Although Aristotle may have described the virtue of practical wisdom as a 'habit of the soul', he appears to have intended that this meant a disposition that could be reflected on by the virtuous person rather than an unthinking behaviour (Hinman 2013: 257–259). Wise social workers therefore cultivate critical reflexivity throughout their careers.

For those social workers who are entering the profession or who are at an early stage in their careers, the cultivation of virtues can be assisted by education, including practical wisdom. Pullen-Sansfaçon (2010) describes how Socratic dialogues might be used in educational programmes to provide a way for social work students to undertake critical reflection and share the insights that each can bring to a particular situation. Pullen-Sansfaçon describes the process in the following way.

> The overall aim of Socratic dialogue is to try to reach consensus on values and principles in a question that is set at the beginning of the process. This question is explored through an analysis of practical experiences. The process includes an exploration of various assumptions and beliefs that underlie common behaviours.
>
> *(2010: 406)*

A dialogue of this kind is facilitated and guided by an experienced social work educator. To ensure that what emerges from the process is helpful to students in developing the critical reflexivity that will be core to their future practical wisdom, Pullen-Sansfaçon (2010: 408) notes that exposure to a range of ethical theories and debates is also essential.

In addition, Banks and Gallagher suggest that 'ethical decision-making models' can be useful in the early stages of learning to be a social worker (2009: 94). At the same time, they emphasise that these are tools for learning and that such models should not be seen as a reason for social workers to avoid the demanding task of gaining experience and practising the virtues. They agree with Papouli, who states, '[p]ractical wisdom can only be acquired through experience' (2018: 6).

Conclusions on practical wisdom

Slote (2001) argues that motives in themselves are not guaranteed to produce good actions. Although agreeing with this basic point, Papouli suggests that conscious attention to how one's qualities and capacities are used, which is practical wisdom, assists in ensuring that social workers may be motivated and thus act 'in the correct way' (2018: 11). What constitutes 'correct' may continue to be a matter of debate among social workers at the higher level of theory (for example, in debates about helping people versus changing social structures), but at the level of involvement in any given situation it becomes easier to see why one action can be agreed as 'better' than another.

Banks and Gallagher conclude their discussion of practical wisdom by noting that one of its key features is the quality of being able to tolerate the messiness of everyday practice (2009: 95). For them, it is wisdom that enables practitioners to make ethical sense of each situation they encounter. Like Papouli, they state that it is this capacity that answers Slote's concern and helps to ensure that the other virtues do not simply become a set of good intentions or, even worse, we may infer, a justification for the habitual and uncritical continuation of poor practice.

In their cross-cultural review of character strengths Peterson and Seligman (2004) locate wisdom at the centre of their short list of virtues that are recognised in many cultures. They point to the qualities that contribute to wisdom as creativity, curiosity, open-mindedness and a love of learning (2004: 29). However, this is not understanding for its own sake but for the purpose of being used well (2004: 39). It is *practical*. In a profession such as social work, whose moral purpose as well as its practical function is to help people to achieve change in both individual lives and social systems and structures, the quality of practical wisdom is at the centre of what it is to be a good practitioner.

Finally, we can observe that practical wisdom is the characteristic by which social workers exercise ethical imagination as well as ethical judgement. That Peterson and Seligman (2004: 110) include creativity as a strength of character points to the way in which the capacity to produce original yet situationally appropriate understanding of the world expresses wisdom. It is for this reason that ethical intuition is valued by many social workers as a means to achieve good practice. In this sense, intuition is the capacity to connect ideas and actions in such a way that the pattern explains what is ethically significant and from that provides a basis for good practice to emerge. Practical wisdom is the quality of being able to see not just to the moral horizon, but beyond it, so that the path to achievement of the goals of practice can be understood.

FINAL REFLECTIVE QUESTIONS/EXERCISES

List three key learnings from this chapter.

Do you acknowledge the 'messiness' of everyday practice in the caring professions, as discussed above? If so, how do you experience and manage this messiness?

If practice wisdom is primarily learned through practice experience, how might this virtue be nurtured and developed in social work students and recent graduates?

How might you further develop your own practice wisdom?

9

INTEGRITY

CHAPTER OBJECTIVES

- To define the virtue of integrity
- To explore the role of integrity in professional social work practice
- To discuss the concepts of promise-keeping, authenticity and honesty in relation to professional integrity
- To consider how social workers can cultivate integrity

The virtue of integrity

As with practical wisdom, integrity is a virtue that is characterised in terms of its relationship to other virtues in the form of the ability to integrate those ethical qualities, to bring them together in such a way that they work together. Banks and Gallagher describe integrity as 'a kind of moral competence or capacity that people use to make sense of their ideals and actions as a whole and to act accordingly' (2009: 195). So although the meaning of this quality as an ethical virtue does not derive from the consequences of the action, we can say that it is only in the actions of people that we can see whether they have integrity. The claimed values of people with integrity match their actions; in everyday terms they 'walk the talk' or 'have a moral compass'.

However, this still does not tell us what integrity's elements are. Which ethical virtues does it bring together? Indeed, are there dispositions or qualities of character that are subsidiary to integrity and which make up its overall appearance? In some accounts, integrity is regarded as a sense of 'wholeness' of the moral character of a person. For example, Williams (1981) considers integrity to be a matter of someone maintaining the life commitments that make up their identity. In a similar vein, Nagel (1979) considers life commitments as part of the set of plural values that informs contemporary ethics. What these philosophers are identifying is the way in which a person can be considered to have moral wholeness when what they do is consistent with their ethical commitments, following through on these as a central aspect of their moral identity. Examples might include an ethical commitment to veganism, to pacifism, or to the abolition of modern forms of slavery, which then informs choices about what to eat, what to buy, how to spend your time, what career to follow and so on. Yet, as Banks and Gallagher note (2009: 201), Williams' (1981) concept of commitment does not differentiate between those life goals that are simply someone sticking to their own assertion of what matters and those that might be regarded more widely as 'good' ends. Nagel's (1979) emphasis on moral commitment is different in this respect, in that he emphasises the ethical dimension of such life projects. For Nagel, we see moral commitments when we can say that a person's choices and actions are 'principled', because they clearly embody an ethical stance. Moreover, this is an ethical position that is consistent, seen across all aspects of a person's life.

Likewise, Cox *et al.* (2003) suggest that in relation to the consistency of life commitments it is more helpful to think of 'moral integrity' as the focus of a person's attention to ethical virtues. In this sense, the person with ethical integrity is one who consistently pursues commitments that bring together 'good' qualities and objectives. The person with integrity not only demonstrates moral commitments that embody ethical values, but also in doing so draws on other moral virtues and combines them into a coherent whole. On this basis, the person who is committed to pursuing veganism, pacifism or the end of modern slavery could be said to be expressing ethical integrity if, in doing so, they also display a combination of ethical virtues that may include courage, compassion, justice, humility and so on.

Writing in the field of education, Musschenga (2001) distinguishes between 'external' and 'internal' consistency and coherence as elements of integrity. The distinction between these two ways of understanding integrity is that someone displays external integrity when their actions match espoused values and goals. For example, Musschenga (2001: 220) suggests that a person with external integrity 'will never abuse the power, privileges and opportunities that [her or] his role provides'. Internal integrity is seen when there is coherence and consistency between beliefs, values and actions. Again, we see here the idea of a person being morally 'of one piece' or a 'whole person' (Musschenga 2001: 221). At the same time, Musschenga asserts, it is not necessary for others to agree with the values being espoused for a person to be seen to have integrity; what matters is whether people 'stick to their promises, act according to the principles they have subscribed to or are expected to act upon in the offices and roles they fulfil, and are loyal to their commitments' (2001: 220). This then leads Musschenga to argue that *moral* integrity does not require someone to have coherence and consistency in their own values, but that their actions be coherent and consistent in relation to the role that they perform. Finally, for a person to be someone of moral integrity they must integrate their own moral self with the ethical values of the relevant public role. In the case of the present discussion, this is the professional role of social workers, in which we may expect their actions and judgements to be coherent and consistent with the ethical expectations of social work.

In the biomedical field Pellegrino and Thomasma (1993) also propose a view of ethical integrity as completeness or wholeness. They expand this approach in a way that connects closely with the discussion of practical wisdom presented in the previous chapter of this book. They say:

> By integrity of the person we mean the right ordering of the parts in relation to the whole, the balance and harmony between the various dimensions of human existence necessary for the healthy functioning of the whole organism [...] expressed in a balanced relationship between the bodily, psychological and intellectual elements of their lives [in which no] element is out of proportion to the others.
>
> *(Pellegrino and Thomasma 1993: 129)*

From this, the integrity of the professional helper is to support and promote the moral flourishing of both the service user and the professional her or himself. At the heart of this is the ethical quality of 'fidelity', that is, faithfulness to the core moral purpose of the profession (whether that be healing, wellbeing or another purpose). This also requires qualities of compassion, commitment and trustworthiness.

Beauchamp and Childress (2013: 40) argue for a more restrictive view of the way in which ethical integrity is understood in the health and human service professions. They say that integrity requires both the wholeness of moral character and being faithful to core ethical values. This view does not only consider someone to lack integrity if they express hypocrisy, insincerity, bad faith or self-deception, but also when the person does not 'act on the correct moral beliefs' that are held (Beauchamp and Childress 2013: 41). From this perspective, ethical integrity is seen not only in what someone does, but also in what they fail to do; not to act well may be as much a failure of ethical character as doing the wrong thing.

Beauchamp and Childress introduce two particular questions into a consideration of ethical integrity. The first is its relationship to the concept of conscience, while the second is the content and extent of the list of subsidiary moral characteristics that can be said to comprise integrity. With regard to conscience, Beauchamp and Childress are addressing the detailed issues that often impact certain areas of health, and which we often encounter in everyday conversation as 'ethical dilemmas', such as abortion, euthanasia and other actions that often attract strongly held moral views. Perhaps less widely recognised as an instance of conscience is when a professional declines to act in a particular way because they consider that what they are being asked to do is futile. Futility can be seen to be wrong because it is a form of dishonesty, it fails to respect the humanity and moral autonomy of the service user and it is very likely to be a wasteful use of scarce resources. Furthermore, although many issues that come under the notion of conscience emphasise those practices in which some health professionals choose not to engage, that is, they entail 'conscientious refusal', there are also acts that a person might feel compelled to undertake on grounds of conscience (Beauchamp and Childress 2013: 43). An example of such a conscientious action can be seen when the professional challenges organisational policy and insists on doing something for a service user 'because it is the right thing to do'. Similarly, a whistleblower can be said to act on the basis of what is right, even though in doing so they consciously do not act on the ethical virtue of loyalty.

However, the range of commitments and actions in professional life that can be guided by conscience in this way is much longer than these well-known examples, even though they are very important to many people. They also raise other questions about respect, rights and autonomy, and justice and fairness, some of which are questions of virtue in themselves and all of which are relevant to the person who wishes to live a life of ethical integrity. This leads us to the range of ethical virtues that may be said to be subsidiary to integrity, even though they are also themselves ethical virtues. We have already noted respectfulness, faithfulness and trustworthiness, to which we can add honesty and truthfulness, sincerity, reliability and loyalty. For Peterson and Seligman (2004: 250) the key qualities that comprise integrity are 'practicing what one preaches', 'public justification of moral convictions, even if those convictions are not popular' and 'treatment of others with care [...] sensitivity to the needs of others'. To explain these qualities they return to concepts of authenticity, honesty, truthfulness, kindness and so on (Peterson and Seligman 2004: 256–257). From this, they note that in some psychological studies measures of honesty are used interchangeably with tests for integrity (2004: 261), which suggests that at least for some analysts the two values are the same. However, towards the end of their discussion Peterson and Seligman summarise integrity as values, goals and actions being congruent and as whether 'actual behavior is consistent with one's morally justifiable principles' (2004: 262). Again, we can note that integrity concerns ethical values and principles for which an account can be given – they are not simply private projects or commitments.

In everyday usage, someone may be referred to as a 'person of integrity' when they display even only some of the qualities – respectfulness, faithfulness, trustworthiness, honesty, truthfulness, sincerity, reliability and loyalty. The point is that they are consistent and coherent in the range of qualities that are displayed. This is the case whether we are considering everyday or 'ordinary' ethics or the approaches to ethics that derive from the particular roles and functions of professional life. Peterson and Seligman (2004: 267–268) also note that integrity is a moral quality that is widely asserted across diverse cultural and religious traditions (compare with Hunter and Kendrick 2009). However, at the same time, the ways in which integrity is displayed can be affected by cultural or religious emphasis on particular qualities, such as respect and politeness or self-effacement, as against individual autonomy and assertiveness. The common dimension that they add to coherence and consistency is that of 'authenticity', which they argue is culturally contextualised and differentiated.

Beauchamp and Childress (2013: 41) are quite clear that with regard to integrity the formal ethics of professions are not necessarily in themselves distinct from society more generally, at least to the extent that they could be used to justify values or actions that do not satisfy the wider community's ethical expectations. However, we can observe that it is hard to find concrete examples of such expectations being lower within a professional context than in the wider community. What is more likely to be the case is that professions have specific issues in which integrity is expected that are not encountered by those outside the profession. Thus, at the very least, someone who is seen by their colleagues as having integrity in a profession is likely to be regarded as such beyond that specific group as well.

Of note for this discussion is that unlike other ways of looking at integrity as an ethical virtue, Peterson and Seligman (2004) locate it as a subsidiary of moral courage. They appear to do this because, following their definition of it as the combination of honesty and authenticity, they regard integrity as requiring the capacity to face opposition, criticism, the loss of valued goals through having to make difficult choices and so on. However, this risks becoming a circular argument, as courage is one of the ethical virtues that integrity combines with other qualities to produce a coherent moral self. As we have seen, from a wider philosophical perspective integrity is widely understood as the ethical quality that disposes a person to bring together and connect their other ethical virtues in a way that makes authenticity, honesty, faithfulness, truthfulness, sincerity, reliability and loyalty possible, and integrates these within a moral self that may comprise courage as well as compassion, care, hope, perseverance, justice, humility and wisdom.

REFLECTIVE QUESTIONS/EXERCISES

Given the exploration of the concept above, how do you define integrity?

What are the differences between 'external', 'internal', 'moral' and 'ethical' integrity, as described above?

Has there been a time in your practice/training where your conscience was activated, that is, you experienced ethical dissonance? What was the situation? How did you act and why?

Integrity as a virtue for social work

Limited attention has been paid to the virtue of integrity in social work ethics until relatively recently (Banks 2010: 2169–2170). Since the beginning of the twenty-first century this has started to change, with a modest growth in concern with this aspect of ethical virtue. Two elements stand out in Banks' discussion of integrity. First is that it is somewhat unlike most of the other virtues, in that it refers to the quality of being able to connect and combine those other qualities. Indeed, this is reflected in the very word, which refers to the *integration* of the virtues. Banks

(2010: 2176) cites Cox and her colleagues' (2003) description of integrity as a 'complex cluster concept'. This is a somewhat more fluid notion than that of commitment that we have discussed above, as it suggests the social worker is carefully balancing ethical virtues in a way that is appropriate for the situation in which they are practising.

Second, Banks notes that professional codes of ethics have only begun to include reference to integrity in this recent period (AASW 2010; BASW 2014). Moreover, such references have tended to focus on the various ethical virtues that integrity seeks to connect, rather than on the quality of integration in itself. This in turn may be a reflection of the extent to which social work has not attended as much to virtue-based as to principle-based ethics (McBeath and Webb 2002). For example, drawing on the Aristotelian tradition, Papouli (2018) makes no mention of integrity as core to the virtues in social work.

Reamer (2018) offers another brief summary of integrity as part of the ethical framing of social work. In general terms he agrees with Banks and the wider professional ethics literature that integrity refers to moral soundness or wholeness. However, to this Reamer adds a more restricted view that integrity refers to 'fidelity in adherence to moral norms' (2018: 52). In other words, the social worker with integrity remains focused on the values of the profession. From this point of view, which parallels Musschenga's (2001) position discussed above, Reamer then suggests that there are two elements of a person's ethical character that are important. The first of these is 'coherent integration of aspects of the self [...] so that each complements and does not frustrate the others', while the second is seen in someone 'being faithful to moral values and standing up in their defence when necessary' (Reamer 2018: 52). To do so requires that the social worker is asserting the ethical values that are formally those of the profession as a whole and not simply being coherent and consistent with their own personal values.

Reamer (2003) also applies this point in a debate with Hodge (2002). Hodge states that social work discriminates (albeit unintentionally), in that both social workers and service users who hold socially conservative religious beliefs are subject frequently to 'hostility' and 'antagonism' (2002: 406, 410). In response, Reamer agrees that the moral integrity of the profession requires that the core social work ethical values of non-judgementalism, self-determination and anti-discrimination should indeed be applied to everyone, whether practitioner or service user (2003: 429). However, at the same time, it also requires that social workers who hold socially conservative moral values grounded in religious faith do not impose their personal convictions on decisions and actions made in relation to service users, nor expect that service users will conform to the social workers' personal ethical values in order to receive assistance. We do not need to agree with another person's values in order to deliver the assistance that our professional role is there to provide. Where the dissonance between personal and professional values is too great, then the social worker with integrity should acknowledge this and seek an outcome that does not compromise the profession's values. For example, it would be reasonable to expect that a social worker who holds conservative views about abortion, based on religious belief or otherwise, would choose not to work in an agency with functions that include providing assistance for women to access abortion services. To choose to take a social work role in such an agency would call into question the moral integrity of the individual in relation to the profession's formally stated ethics, especially if this was done with the explicit goal of subverting the agency's functions.

Banks and Gallagher (2009) provide the fullest account of integrity as it affects social work (and related professions such as nursing and allied health). For them, integrity in social work is to be understood as a combination of 'ordinary' and 'special' ethics (2009: 204–205). First, they argue that we should consider the way in which personal integrity is displayed in professional life. Drawing on the work of Musschenga (2001), Banks and Gallagher identify a 'global' approach, by which they mean a coherent framing of values and actions across all aspects of a person's life, which is then applied to the 'local' context of that person's professional role. In this sense, integrity concerns the way in which the everyday moral self of the practitioner is then manifested in the role they perform in the specialised circumstances of professional action. Second, there is the distinct value of 'professional integrity' in which qualities such as honesty and reliability are framed specifically in the 'local' contexts of the professional role, based on the shared values of the profession. From this, Banks and Gallagher (2009: 205) argue that both personal and professional integrity are important in professional life, because although the professional role may modify personal values and qualities, at the same time professional values and qualities have to be incorporated by the practitioner, who may revise them in some way in how they are accepted and applied.

Previously we noted Musschenga's distinction between 'external' and 'internal' coherence and consistency in integrity. In relation specifically to professional identity and practice, this distinction raises the question about whether there must be coherence and consistency between the external and internal domains (Musschenga 2001: 221). In other words, how much should we be concerned with the relationship between a person's values and actions in their private life and in their professional role? For example, should we challenge the integrity of a social worker who brings together appropriate values and qualities in their work but is found not to have thought and acted in the same way in other areas of their life? Banks and Gallagher recognise that for some social workers, the strength and focus of their own moral commitments is such that if they are not able to connect the personal and the professional they will consider that their integrity is compromised. For example, they quote a social worker who left a job because he was not able to practise in a way that was consistent either with his own values or how he understood social work's professional ethics (Banks and Gallagher 2009: 197, 206). In contrast, for others the concern of professional ethics is solely on whether there is sufficient integrity between acceptance of values and qualities defined by the profession and their actions in performing their professional roles. From this second position, whether or not you hold moral commitments that are more demanding in the rest of your life, some compromises may need to be made in the circumstances of practice. For example, resources may be limited, or the range of options in a particular situation may be limited to a choice between actions and outcomes that are equally 'bad'. The difference between these two positions can be summarised as whether or not integrity is unacceptably compromised in having to accept and work with contextualised compromises.

To summarise their arguments, Banks and Gallagher list six characteristics of integrity that are common to the view that both personal and professional ethics must be coherently brought together and the view that we should only be concerned with professional ethics. These are:

- a commitment to a set of values that define the 'good professional';
- an awareness that values are inter-related and form a coherent whole;
- a capacity to make sense of professional values;
- the ability to give a coherent account of values and actions;
- a disposition to think, feel and act in accordance with core values;
- strength of purpose and the ability to implement these values.

(Banks and Gallagher 2009: 209)

So, based on this set of criteria, the social worker of integrity is not simply someone who displays the values and qualities that are defined in professional codes of ethics. They are able to be 'morally active' (Husband 1995) in taking responsibility for making the connections between values and qualities and integrating this in their professional identity and actions. This social worker is sufficiently ethically self-aware to be able to exercise critical reflection and to be accountable.

REFLECTIVE QUESTIONS/EXERCISES

How do you define integrity in a professional social work context?
Are there differences between personal integrity and professional integrity?
What do you think a 'morally active' practitioner looks like in practice?

Integrity in social work practice

In Chapter 3 we discussed the ethical virtue of courage and used two hypothetical vignettes of practice in which social workers sought to challenge the decisions and actions of other people. 'Jane' is described as helping a family seeking asylum to oppose the government department responsible for making decisions about their status as refugees; 'Gloria' is seen working with a family where the use of physical punishment of children is regarded as abusive. In Chapter 4 we looked at another hypothetical vignette, of 'Cheryl', in which a service user appears to undermine the help she has been given, while in response the social worker demonstrates compassion. In each of these situations we portrayed social workers who thought carefully about the connection between the values and goals of their work and who also brought together their personal commitments and the professional ethics of social work. Yet there is a difference, in that Jane's work involves opposition to policies established by the elected government, whereas Gloria and Cheryl are implementing policy within the prevailing legal framework. In each instance the social worker is seen as coherent and consistent in their connection of the virtues of courage and compassion to the rights and needs of the service users, in their carefully considered use of professional knowledge and authority and their respectfulness towards the different service users who are impacted by their work. In each case the social worker also displays a personal commitment to the ethical values and the goals of their practice.

Among the social workers with whom we spoke, their expressions of integrity can be distinguished in three ways: as promise-keeping and taking responsibility; in terms of 'being real' in the sense of being able to be true to your own character; and through 'authenticity', defined as being able to connect personal and professional ethical values. Together these qualities contribute to the sense of 'walking the talk' that some use to summarise the ethical virtue of integrity.

Promise-keeping and taking responsibility

Some approaches to virtue ethics consider promise-keeping to be an aspect of distinct virtues, including fidelity, respect, trustworthiness, honesty and perseverance (Pellegrino and Thomasma 1993; Beauchamp and Childress 2013). Social workers regard keeping promises or commitments, taking responsibility and being honest as part of their integrity. For example, when asked to illustrate a comment about integrity, a mental health social worker described the importance of keeping promises and taking responsibility by being open and honest as central.

> So calling her, when you said you were going to call her. And if you didn't, still calling her and apologising for that [...]. Following through on things, so you know, being really clear from the outset and if I'm worried about your safety, I don't really care if you agree, I have to call somebody about that. And being really clear about who you are going to call, and who you're going to get involved.

For this social worker integrity is seen in the way that doing what she said she would do, combined with honesty and openness, enables her then to bring together personal and professional values. Although the outcomes of practice do not determine the rightness of these actions, in this instance the practitioner also sees this quality as important in practice that the service user saw as good.

Similarly, a senior social worker in a community organisation connected promise-keeping with honesty.

> I think what we did well was to be honest, to be very inclusive in the decision-making, to share the information. There was no reason not to tell

> everybody everything. And to be very open about that, so when I met with the carers I could talk very openly about the situation and why we were in the situation we were. But too, I think to hold out the promise that while things were bad or dire at this point there was no reason why things couldn't get better, so I think that was a good thing to do.

This social worker also makes the point that ethical promise-keeping requires honesty and openness in *not* making promises that are known in advance to be difficult or impossible to keep. While this may appear to be an obvious corollary, it is also possible that when a social worker makes a statement to encourage a service user to have hope, that it is instead heard as a promise. In this way, practitioners need to be careful to ensure that an appropriate balance is sought between instilling hope and being honest, so that integrity reflects practical wisdom.

Taking responsibility by being open and honest so that promises are only made when they can be honoured is expressed by a social worker in describing practice with a particular family.

> I think what we were, and [...] feedback from one of the families was, 'You were always really clear about what this journey would be like,' because it turned to be for this particular family, a pretty tortuous journey, and probably goes on to be for times a bit tortuous. So I think that we were really transparent with the people engaged in this new process, and [...] that's just what you know to be good practice, but we clearly did that well.

Here again we see how honesty is connected with respectfulness, so that the family were able to recognise this and not to feel afterwards that they had been misled about the difficulties of facing the issues they needed to address.

Taken together, these examples illustrate the way in which promise-keeping is related to taking responsibility and in doing so combines other ethical virtues in practice.

Being real – being yourself

A somewhat different expression of honesty is seen in the way that some social workers place importance on 'being real', in the sense of being true to their own values. This can at times be difficult, as a social worker in a non-government case management agency noted, recalling a service user who asked:

> 'So who do you think reported me?' and because of the way that we'd been working I just shared honestly with her 'I was really concerned about you because I couldn't get a hold of you. I made a report to [name of agency],' anyway not long after she disengaged and [...] I really wondered if I'd done the right thing, if I shouldn't have said anything to her [...] I needed some reassurance from the team. Did I do the wrong thing? [A colleague from]

> another service commented to me 'Oh you told her you made a report?' Like yes, did I do the wrong thing? But that's kind of the way we're working a little bit in [name of agency] is that we often say to our clients 'If there's something, if I'm really concerned about something that I'm seeing with the kids, before I do anything with that I want to talk to you about it' [...]. And maybe, what would have happened if I didn't tell her that I made a report?

The reaction of the service user caused this practitioner to question her own judgement, but on reflecting about the conflicting values involved she was able to reach a position that integrated the underlying professional values with her decision.

In a somewhat different context, community mental health, another social worker described being able to act on his own values in a way that connected honesty, trustworthiness and respectfulness.

> I had one guy who was so traumatised by the system, and clinicians, so my whole focus for him was to fix up the systemic flaws to ensure that he felt comfortable as much as he could [...] it took me literally three weeks to sort all this mess out [...] and by doing that, he then assessed that I was okay and I was trustworthy [...] I think that's the key, for me [...] that they will feel that they'll be heard, and what they, it's about involving them, really involving them in their care and what's happening to them. It's not this tick and flick and this tokenistic 'oh, I'm listening to you.'

So by taking such an approach this practitioner was able to be true to his own perspective that all aspects of the way in which a service user is treated should express recognition of their humanity through the embodiment of his own values.

A social worker in a counselling role is even more explicit that having a job that is congruent with her own ethical perspective is important, when she says of the agency and her own values:

> I think they are closely aligned and that's why I chose social work because for me, it's very important to live your values and I saw social work gave me an opportunity to do that but I think there's a lot of conflict there as well sometimes [...] because I'm an idealist I get very frustrated by certain [...] I'm not able to work with people in the way that I would ideally want to so I get quite frustrated with that or [...] by what I perceive is kind of contradictions in the practice.

This statement conveys a sense that although in one respect this practitioner sees the profession as a place for expressing her own values, so that in broad terms she can achieve personal integrity, at the same time specific aspects of the role she fulfils make this more difficult. It is not always possible for her to realise her values because of limitations and contradictions in the day-to-day work.

This contrasts with a practitioner who states that there was 'zero integrity' in the agency in which he was working until recently. This was a significant factor in him choosing to leave that organisation. He comments:

> I find that very difficult, because the level of integrity that I live with, always everything is about the quality of people. Doesn't matter if you're a client, a staff member, if you're a stakeholder, that same level of care, love and respect should be offered. You know, those mission statements should be lived and not just written on a piece of paper to look good [...] like we have this idea that we have separate lives, you know, I can be this person here and I can be that person here. Where for me I feel that that doesn't hold integrity [...] the way that I live in my personal life then dictates the quality that I bring to my [work].

Similar to the example of youth worker 'George' discussed by Banks and Gallagher (2009: 197), this practitioner eventually decided that he was not able to remain within a role that required him to compromise on the importance of his values and actions fitting together in an ethical whole.

Authenticity – connecting the personal and the professional

The distinction between 'being real' and 'authenticity' is subtle, as there is some overlap. Linguistically, these terms convey a very similar meaning. However, for social workers there are two elements that are important here – one is whether you can be true to your (ethical) self and the other is whether there is coherence and consistency between your own values, those of the profession and those of the agency or role in which you work. The social worker quoted above, who is critical of 'zero integrity' in his former place of employment, partly blends the two, while emphasising the need to maintain his own integrity. However, the following example reflects a social worker who looks at how her values can be coherently and consistently connected with those of others. She says:

> Authenticity is so important. I value it in my work with other people. I've had situations where I've thought, 'I won't work with that person, I just won't. They make things worse and I'm not prepared to engage with that kind of crap' [...] if there's no authenticity and integrity I don't work with it. I might have to weave my way around that very delicately, but there's an agency at the moment that I have nothing to do with because I think their work is damaging and dangerous [...] authenticity and integrity is really, really important to me, and at the absolute heart of the corridor conversations that I have with key players in different spaces. It's the relationship that is fundamentally important.

So from this perspective integrity is understood as the way in which different sets of values can work together, supporting each other while also connecting personal values with the wider values of the profession.

A similar view is also expressed by a social work manager who notes that:

> probably [the best] advice I'd give other social workers is to be authentic. Don't go into a session pretending that you're a particular kind of practitioner; we're all different practitioners. [However,] it's a bit like I said before about the professional and the personal. You got to put on that professional role, realise that this is a job that you've taken on, it's your responsibility to deal with it. And your job is to do the best you can with the information and the resources that you've got, and that's your job to do. You don't duck the issue.

From this perspective, integrity is to be found in a balance between the authenticity of 'being real' (in other words, true to your own values) and that of taking responsibility for the professional role that brings you to the encounter with service users, in which the struggle for coherence and consistency of values occurs. Such a balance may appear to be a juggling act at times, as the social worker seeks to make sense of the way in which personal and professional ethics can be achieved without there being an unacceptable compromise in either direction.

Social workers who leave their job because they are unable to 'be real' ethically are demonstrating integrity. However, given the limitations of the social, political and economic circumstances in which social work is practised, for most social workers finding integrity between personal and professional morality is a critical, everyday challenge. It is not only very difficult – it may even be detrimental – to separate our personal moral identity from our professional ethics (Oakley and Cocking 2001: 138), but holding them in a creative and constructive tension may be a way that many social workers achieve integrity, while for others finding a congruent fit is the only way that they can achieve this ethical quality.

REFLECTIVE QUESTIONS/EXERCISES

How might we balance hope and honesty in social work?

Do you think promise-keeping is important in social work? Why/why not? What does promise-keeping look like in practice?

How do your personal values influence your work? Is there consistency between your personal and professional values, or are they separate? Why?

Cultivating integrity

This discussion of ethical integrity in social work has pointed to a complex mixture of personal and professional factors in the ways that practitioners make sense of what this quality requires and its importance in their identity, as well as in how they fulfil their roles. As with practical wisdom, this virtue is one that clearly requires conscious attention and practice. Although some social workers bring an innate capacity for integrity to their work, others may need to acquire both an understanding and the ability to pursue this virtue. Gough and Spencer (2014: 31) note that achieving integrity between personal and professional moral values requires a capacity to cope with the complexity of different values that are encountered in practice, as well as the reasonable diversity of ethical positions and differences between the values of individual social workers and between social workers, service users and members of other professions. Being able to hold onto your own value position while accepting that others may be different rather than wrong, and when different *is* wrong, is a considerable ethical quality that forms a core part of integrity.

Banks and Gallagher (2009: 210–211) suggest two areas in which development of ethical integrity can be promoted. The first is in formal education and the second is in the community of the profession, through organisations and networks. In formal education they argue for three components: debate about professional values, perhaps through exercises to explore ethical dilemmas; teaching about the history and development of social work, so that the changing context and nature of the profession and how this shapes ethical values is understood; and practising respectful dialogues and debates so that differences can be explored. Banks and Gallagher suggest that, after qualifying, continuing collective discussion of professional values through professional associations, trade unions and similar organisations can assist social workers to strengthen their ethical integrity, including provision of solidarity in settings and relationships in which serious ethical challenges can be discussed, such as pressure to tolerate or cover up ethical problems.

All of these proposals provide ways for ethical integrity to be developed and sustained. Although she does not address the quality of integrity explicitly, Pullen-Sansfaçon's (2010) model of Socratic dialogues in social work education is one possible way of constructing such a strategy. Its contribution could be found in the way that it canvasses different points of view while at the same time seeking at the end to move towards a shared perspective. Thus it models the relationship between personal and professional moral values.

Papouli (2018) likewise does not explicitly address ethical integrity; however, she points to the Aristotelian notion that ethics concerns life-long learning. From this perspective integrity, as much as any of the other virtues that it seeks to bring together, may be considered a 'habit of the soul'. This makes the cultivation of ethical integrity a life-long project that is both the responsibility of each social worker and also of professional collective groups. For that reason, each social worker shares the responsibility for modelling and encouraging colleagues in developing and maintaining integrity, especially when holding a supervisory role. With regard to integrity, this could even be considered part of the overall framework of professional ethics. For example, section 3.3 of the Australian Association of Social Workers' Code of Ethics (AASW 2010: 13) calls on members of the profession to 'ensure ongoing professional competence by participating in and contributing to their own life-long learning, education, training and supervision, and that of other social work practitioners and students'. Insofar as ethical integrity provides a way of bringing other values and qualities together, contributing to the development and sustaining of this capacity by colleagues as well as yourself might be regarded as part of ethical practice in itself.

Conclusions on integrity

As Oakley and Cocking express it (2001: 139) integrity is achieved when all aspects of who we are and the profession in which we are engaged 'speak with the same voice'. For those social workers who can say that their own values are coherent and consistent, and that they are also closely aligned with the values of the profession and the role that they fulfil, ethical integrity provides a sound basis for practice. Those whose values are challenged by their role and work context, or who experience ethical dissonance in their work, face a choice between seeing organisational or policy change as part of their professional goal or seeking work in a setting that enables them to achieve integrity. Banks and Gallagher's (2009: 210) point about understanding the history of the profession is helpful because sometimes the past is seen more positively than it might have been, whereas in many ways social work has always had to confront unjust policies and practices, limited resources and a lack of support in the wider community (compare with Payne 2005). In such circumstances ethical integrity centres on maintaining the coherence and consistency of a person's own values and those of the profession and using those as a basis for critical practice as far as is possible.

Although it is not widely addressed in discussions of social work ethics and values, integrity is a central quality that should be recognised – along with practical wisdom – as offering a framework for making sense of and drawing together the other ethical virtues that support and sustain good social work practice.

FINAL REFLECTIVE QUESTIONS/EXERCISES

List three key learnings from this chapter.
How might integrity be learned and cultivated in social work?

10

ETHICAL VIRTUES IN SOCIAL WORK – TOWARDS VIRTUOUS PRACTITIONERS?

CHAPTER OBJECTIVES

- To summarise the content of the preceding chapters
- To draw attention to the importance of context to virtue ethics, exploring how different virtues are valued across different times, fields of practice and social and cultural contexts
- To present concluding arguments on the significance and contribution of virtue ethics to social work

The virtuous social worker

Throughout the preceding chapters we have considered various central elements in the ethical character demonstrated by social workers. In drawing these discussions together, this final chapter returns to overarching issues that are identified in Chapters 1 and 2 in order to make connections more explicitly between the virtues that have been explored in depth by Chapters 3 to 9. In particular, it is important that we conclude by addressing the relationship between these virtues and the way in which together they contribute to good social work practice through the formation of virtuous practitioners.

In summary, Chapter 1 examines the approach of virtue ethics while Chapter 2 discusses the application of the concept to applied and professional fields. Central to this analysis is the way in which virtue ethics shifts our attention away from formal principles, which are usually held to exist independently of those who enact or are affected by them, to the moral character of social workers themselves. We conclude that virtue ethics arises from the community of practice of a profession.

This in turn opens up the possibility that contextual differences can reasonably affect the way that particular virtues might be emphasised in particular settings (*these* virtues, for *this* practice *here*). Notably, the way in which culture affects the understanding of ethical virtues in various settings is more readily addressed by such an approach. In addition, a virtue ethics approach enables facets of human existence such as emotion and relationship to be seen as contributing towards ethical understanding, in addition to the contribution of rational thought that is central to principle-based ethics. This creates a more holistic basis for professional ethics. However, we also note that there continues to be relatively little attention paid to virtue ethics in social work (notwithstanding the exhortations of McBeath and Webb [2002] and others). The continuing emphasis on principle-based ethics can be seen as a consequence of the importance of formalised codes of ethics in a neo-liberal world, in which risk management and accountability are widely seen in terms of prescriptive, impersonal, bureaucratic, legalistic and technical rationality rather than indeterminate and nuanced human relationships (compare with Bauman 1994). In principle-based ethics the possibility of 'moral luck' is removed, whereas in virtue ethics the effect of circumstances on ethics is both recognised and accepted (Williams 1981; Berlin 1992; McBeath and Webb 2002; Houston 2003, 2011).

Social workers who have examined virtue ethics for the profession have largely agreed that one of its strengths is that it enables us to consider particularities between specific contexts while at the same time looking at social work ethics within an overall framework (McBeath and Webb 2002; Houston 2003; Banks and Gallagher 2009; Pullen-Sansfaçon 2010; Banks 2010; Papouli 2018). Different practices undertaken in different settings may call for particular moral qualities to be emphasised. This affects the way that ethical virtues might be demonstrated by the good social worker in that context. Furthermore, it is also the case that two equally virtuous social workers might bring different qualities to the same situation (Banks and Gallagher 2009: 218). Yet, such difference is also largely seen in terms of emphasis: this is not an argument for an individualistic understanding. All contributors to the development of virtue ethics in social work base their discussion in the view that such virtues are established and supported by a community of practice.

Arising from this broad agreement we can also see a continuing wish to try to identify the key set of ethical virtues that might be expected of social workers across the profession. Banks and Gallagher provide a useful summary of the ethical virtues identified by seven key philosophical contributions and from eight applied and professional ethicists (2009: 224–225). The most commonly occurring virtue, present in 13 out of 15 of the texts they review, is (moral) courage; the next two most widely identified are temperance (nine out of 15) and justice (seven out of 15). However, no one virtue is discussed in all of these instances. Similarly, in Chapter 2 we observe that the discussions of virtue ethics in social work have largely avoided specifying particular virtues. They have discussed the idea of virtue ethics more generally (McBeath and Webb 2002; Houston 2003; Adams 2009; Pullen-Sansfaçon 2010; Webb 2010), or else have focused on one specific virtue as their core concern (Koenig and Spano 2007; Banks 2010; Donaldson and Mayer 2014; Hay 2019). Only Bowles and colleagues (2006), Banks (2012) and Papouli (2018) have offered detailed lists. However, Bowles and colleagues' list is a combination of moral virtues and other qualities, Banks (2012: 74) describes her list as 'partial', while Papouli (2018) focuses on Aristotle's four 'cardinal' virtues applied to social work without considering other possible relevant virtues. None of these discussions argue that their list is exhaustive or complete.

As with Banks and Gallagher (2009), our own conversations with social workers have demonstrated that there are some moral characteristics that social workers in particular social and cultural contexts share sufficiently for these to be addressed as a reflection of the ethical climate in the profession in those particular situations. At the same time, like Banks and Gallagher, the ideas that we can draw from practitioners cannot be seen as prescriptive. Although these definitions of virtues tell us something important about the ethics of social work, there may be other virtues that have not been identified here; or there may be other possible ways that these virtues are understood and the relationships between them. For example, our analysis reinforces the conclusion that courage is a moral quality that is regarded as important in social work (compare with Rhodes 1986; Bowles *et al.* 2006; Banks and Gallagher 2009; Pawar *et al.* 2017; Papouli 2018). At the same time, although some practitioners might regard courage as a (even *the*) central virtue, other equally virtuous practitioners may reasonably have a different set of ethical priorities that leads to different virtues being emphasised, despite there being an agreement between all these practitioners that the needs and interests of service users are a priority.

For these reasons, the ordering of the virtues that are discussed in this book does not convey any sense of priority between the relevant characteristics, nor does it reflect any quantitative occurrence in the conversations with practitioners that have been drawn on. We started with courage in Chapter 3 because it is the most widely discussed ethical virtue in social work, and Chapters 8 (practical wisdom) and 9 (integrity) address moral qualities of practitioners and the capacity to organise and prioritise other virtues in practice (virtues that have both ethical and technical dimensions). However, this does not imply that any one virtue has greater importance than others; in different situations various virtues are more or less relevant. Each of the virtues explored in this book describes ethical qualities and characteristics that are relevant for social workers in diverse ways.

REFLECTIVE QUESTIONS/EXERCISES

Which virtues do you think are most important in social work practice, and why?

Are there any virtues not discussed in this book that you feel are important in your practice?

In this concluding chapter we are brought back to the distinctive nature of virtue ethics, namely that it considers who people are (answering the question 'what sort of social worker should I be?') rather than how they act (it does not answer the question 'what should I do?'). In this sense, virtue ethics is focused on 'being a good person', which in professional terms can be understood as 'being a good social worker', not simply in relation to technique (such as in 'conducting an effective interview') but with regard to the personal ethical qualities that we explore in this book. Awareness of this is something that is not easy for many social workers to acknowledge or discuss. When asked directly about 'virtues', for example, most (80 per cent) of the social workers with whom we spoke found it difficult to apply the concept to themselves (even though they had volunteered to take part in research about virtue ethics), but gave clear accounts of how they envisaged 'the good social worker' in the abstract. One social worker even questioned the word 'virtue', arguing that in everyday life it suggests someone who thinks well of her or himself, or else someone who is judged by others always to do everything correctly. So, ironically, the idea of 'virtue' was seen in these terms as contrary to the virtue of humility as discussed in Chapter 7. Yet, when this practitioner spelled out her understanding of ethics, it described the core of virtue as being: 'about the person's values and beliefs […] that helps them to be a good social worker'.

In Chapter 1 we also discuss the problem of the circularity in the approach to virtues that defines them in terms of the qualities of character of the virtuous person (Houston 2003) and the possibility that this can be resolved through a dialogue among a community of practice (compare MacIntyre 1985, Slote 1992 and Hursthouse 1999 with Houston 2003). So, that there is commonality between the virtues of social workers presented in this book and the study that is discussed by Banks and Gallagher (2009) and Banks (2012) points to the potential for these qualities to be worth considering sufficiently robust to inform further debate about 'good social work'. It is not that virtue ethics has to fall back on generalised principles at this point, but that virtue ethics has greater capacity to inform the detail of professional life by responding to the specifics and the nuances of everyday situations. (We discuss this question further below.)

We also acknowledge that these views of social work come largely from very similar social and cultural contexts. In Chapter 1 we recognise that the dominant approach to virtue ethics, as to professional ethics more widely, comes from the European ethics tradition and that, as a consequence, further consideration should be given to African, Asian and Indigenous ethics, even though there are some comparisons that can be drawn and interpretations made. Thinking about ethics in relation to a person's moral identity may also provide a basis for cross-cultural ethical developments, as notions of virtues applying to people exist in many parts of the world in a way that concepts of rights or some other ethical principles do not (Hugman 2013). Such a possibility also offers an opportunity for a more genuine dialogue between different parts of the word than the tendency for global Northern ideas to be imposed on the global South, which some have argued is seen in other aspects of social work in addition to ethics (Yip 2004; Gray 2005). As yet there remains little in the Australian context connecting Indigenous ethics with the European-derived frameworks that predominate. So although this is a potentially fruitful goal for social work, it remains to be done.

> **REFLECTIVE QUESTIONS/EXERCISES**
>
> What are the differences between the principle-based and virtue ethics approaches to social work practice?
> What are the strengths of the virtue ethics approach?
> Why is the 'community of practice' vital to the development of virtue ethics?

Virtues in context

Not only are various ethical virtues emphasised differently between practitioners, there is also some evidence from the conversations with practitioners that have inspired this discussion, and from other work undertaken by the present authors (Pawar *et al.* 2017, 2019), as well as the other studies to which we have referred, that a virtue ethics approach enables us to see variations in emphasis between fields of practice and even between specific instances of practice.

This suggests that there is no one single 'right way' to engage in social work. A practitioner might reasonably demonstrate various aspects of their ethical character in responding to different situations. Moreover, it also suggests that there is no one single type of person who is suited to social work. We agree with Davies (1994) that it is possible to suggest some qualities that indicate a person may not be suited to social work. For example, the person may hold a conviction that they know 'the one right way' that someone else can resolve their problems because the social worker has resolved similar problems that way, which we can see as contrary to ethical humility. However, ways of demonstrating the quality of humility are diverse and dependent on setting, the people involved, and other contingent factors. Although this might suggest that a person without any of these virtues would be unsuitable for a career in social work (compare Keith-Lucas 1972 with Davies 1994), even that may be an overstatement, as both the evidence of our own conversations with social workers and that of Banks and Gallagher's (2009) study point to the ways in which sometimes people find qualities in themselves that they were less aware of and which only became evident in the context of a practice situation. Virtues can grow and develop and require nurturing. Only when the lack of ethical virtues becomes clear as an indicator of continuing poor practice might the approach of using virtue ethics to evaluate someone's appropriateness for social work have any bearing. However, the person who lacks any moral courage, who is unconcerned with justice, who has no hope or perseverance, and who is arrogant, is unlikely to fit comfortably with social work – and even if drawn to it for other reasons will find it challenging. Thus they might well make the decision not to stay in social work, without it having to be others who make such a judgement.

An example can be seen in the virtue of moral courage, evidenced when social workers are inclined to face opposition to the pursuit of good ends through appropriate means, as discussed in Chapter 3. However, even this quality can be seen in different ways in various fields and methods of practice. The caseworker in a government social service agency can demonstrate courage in how they seek social justice with and for service users, as much as the community worker in a non-government agency (O'Brien 2011); the mental health social worker may understand courage differently again, as we also saw in our conversations with practitioners discussed in Chapter 3. There is also a risk of a quality such as courage being seen as important in a way that implies a 'heroic' image of the social worker (Marston and McDonald 2012). More often its combination with other qualities such as perseverance (Chapter 5), humility (Chapter 7) and integrity (Chapter 9) support courage and make it possible. Recognising the relevance of less widely known virtues helps to avoid the distortion that a 'heroic' image might create. While moral courage is widely regarded as a social work quality, it does not have to be evident in the same ways or to the same extent in all social workers or in all situations.

This point reconnects our discussion of ethical virtues demonstrated by social workers to the purpose or function of social work. Social workers often choose to work in particular types of agencies or in fields and methods of practice that they recognise suit their character. This is evident in the discussion of integrity in Chapter 9, notably when practitioners talked about 'keeping it real' or 'being authentic', which referred not only to being genuine and honest, but also to being able to work in a setting or in a field that enabled them to be themselves. It is not that the organisational factors render social workers unable to demonstrate virtues such as courage, care, justice, honesty and so on, but that structures shape opportunities and constraints around individual practice. The practitioners with whom we spoke were aware of these dynamics. Although some saw this as a matter of organisational hierarchies getting in the way of good professional practice, there was also a view that social workers can use their ethical qualities to choose how to interact with organisational opportunities and constraints.

Organisational issues are compounded for those who consider 'real' social work only to be when working directly with service users in person. For example, one health social worker complained that her role was now split '70/30 [with] 70 per cent on the computer, or phone calls, referrals, case notes, making the case plan, completing all that, it kind of feels like that [...] things like that really do distract from [work]'. Another practitioner in child protection estimated the split of this kind in her role as 80/20. Yet both also talked about creating spaces in their work to engage with service users, and in the way that they discussed this they demonstrated ethical virtues such as care, compassion, hope, justice, humility, practical wisdom and integrity, and both gave clear examples of how this looked in their interactions with service users. Yet others gave examples of how they have engaged with the politics of social services in a neo-liberal environment, advocating against cuts in practice positions and in the provision available to the public. Whether or not these social workers felt able to challenge the system, work within it for change, work within to create a space for positive practice or be completely constrained appears to be partly related to the qualities that they describe and demonstrate. This in turn is influenced by whether or not these social workers separate ethics from politics or if they connect them. For those who see ethics and politics as connected, the challenge of ensuring that good social work can be practised is not considered an anomaly in the way that it is for those who see ethics and politics as two separate sets of issues.

REFLECTIVE QUESTIONS/EXERCISES

What do you consider to be your 'real' work as a social worker? Why?

Reflect on your choice of field of practice/social work role (or, if you are a student, the preferred area you would like to work in). What influenced your choice?

Virtues changing and developing

Ethical virtues are not static. In applied and professional ethics there are two ways in which the changing nature of virtues must be grasped. The wider identification of specific moral qualities as virtues has developed over time as social work has changed. Several of the studies of particular social work virtues that we have already noted state that the characteristic with which they are concerned has altered, either growing or diminishing. For example, Koenig and Spano (2007: 47) consider hope to be an established social work virtue, in that they assert it has long been seen as a determining factor in the capacity of social workers to promote change on the part of service users. At the same time, they argue that this quality needs to be learned by practitioners, with the implication that hope may grow or diminish according to whether or not it is cultivated. They also recognise that organisational and political constraints can challenge hope (Koenig and Spano 2007: 50–51). Similarly, Hay (2019: 365) suggests that care has been an important ethical characteristic in social work 'since its inception', although it has been 'taken for granted' and only been recognised in its own right in recent years. This is an account of a quality that has changed in its visibility rather than its presence, up to the present when it is under threat from neo-liberal economic and political pressures. (Here too we see an account that connects ethics and politics.) Again, Banks (2010: 2171) states that integrity has been focused on only relatively little in applied and professional ethics, including in social work. However, she points to the ways in which integrity has become much more prominent as some codes of ethics have included a section on 'professional ethics' in which virtues are claimed alongside the principles that are set out in other sections, so that in recent years it has become a more explicit issue.

These three accounts of particular virtues share a sense that ethical virtues are dynamic as well as contextual. For this reason we should not see the growth or diminishing of attention to any virtue as inevitable, but rather as the result of choices and actions influenced by different factors that reflect the views and priorities of different groups. Likewise, for individual social workers the presence of particular virtues, or the degree to which they are demonstrated, is dynamic rather than static. The social worker who previously paid little attention to courage, or to hope or to humility, may come to see it as more important than before. In the examples from conversations with practitioners discussed in the relevant chapters we can see that practitioners change their views about the core qualities that they bring to social work, or that the profession requires of them, and that as a consequence they consciously pay more attention to the ethical virtues that have meaning for them at the present time. An example is seen in the social worker in Chapter 4 who spoke about compassion in working with the perpetrators of domestic violence being expressed through challenging them about their actions as a quality that is developed through both training and experience. Similarly, the practitioner in Chapter 6 who wanted to regain a stronger explicit sense of justice, or the person in Chapter 8 who described how she had developed practical wisdom from a recognition that although emotion is a barometer of values it requires conscious thought to enable that to inform who one is as a social worker, both demonstrate the way in which understanding and application of ethical virtues can change, sometimes consciously so.

Throughout our discussion we have made observations about the cultivation of each virtue. Although some social workers may innately possess certain ethical qualities, even if that is the case they can be developed and grow. At the same time, we can also expect that for many social workers there are virtues for which they have the capacity but which have not been developed in any way. The social worker turning her emotional response into practical wisdom through conscious attention is an example of the latter. There are several suggestions from those social work writers who have addressed this issue. Some propose the formal inclusion of virtue ethics education in social work professional degree programmes (Koenig and Spano 2007; Banks and Gallagher 2009; Pullen-Sansfaçon 2010; Flesakar and Larsen 2012; Donaldson and Mayer 2014; Papouli 2018); this is a tradition that goes back to Aristotle. However, our own review of ethics teaching in Australian social work degrees (Pawar *et al.* 2019) suggests that in the early twenty-first century, at least in formal classes, virtues are not addressed very much.

However, all of these writers and others (McBeath and Webb 2002; Clark 2006; Adams 2009; Banks 2010) also refer to the other aspect of the Aristotelian approach that moral virtues are 'habits of the soul', with the implication that habits must be developed and sustained through repeated practice. This can include practical elements of formal experiential learning, such as the field education placements, as well as learning through reflection on your own practice and drawing on feedback from service users, supervisors, mentors and other colleagues, as well as the influence of professional bodies and further study. This was the way that many of those social workers with whom we spoke described developing the ethical qualities that they discussed with us. Finally, identifying individuals who stood out as good social workers and learning by example was also referred to, although not as often. The concept of the 'moral exemplar' has a long tradition in virtue ethics, and was the basis of the linked study that we undertook examining the biographies of leading Australian social workers (Pawar *et al.* 2017). Being able to gain from such people is an illustration of 'moral luck', because it is entirely circumstantial whether or not such a person is encountered.

Looking at the insights from these published studies and from the experiences of social workers, we can conclude that often ethical virtues are learned and developed through a combination of formal education, reflection on experience and opportune encounters with 'moral exemplars', rather than one or the other. Some eventualities cannot be planned for. Yet it is also possible to be prepared to make use of those opportunities and experiences that are available, and it is certainly the case that each social worker must be willing to engage with such chances when they become available. That is something that education, mentoring and modelling can deliberately seek to encourage. Greater attention to virtue ethics in formal professional education would, therefore, provide a foundation for social workers to take responsibility for their own continuing professional development, through critical reflection (Gardner 2014; Pawar and Anscombe 2015).

REFLECTIVE QUESTIONS/EXERCISES

Which virtues do you feel are most important in present-day social work? Why?

Have the virtues or traits that you consider important to social work changed over time? If so, in what ways and why?

Virtues in a pluralist ethical perspective

In moral philosophy the assertion of one approach as predominant is known as 'monism' or 'absolutism' (Hinman 2013). Examples of this stance include the assertion that only deontology matters (such as the ethics of duty to principle such as human rights), or that only a consequential view is correct (such as the pursuit of equality, as the right way to see social justice). Its opposite is usually held to be 'ethical relativism', which is the position that holds all ethical views to be matters of individual experiences and shifting values. A version of this is 'cultural relativism', in which values are seen to be different between cultures in such a way that there is no way to make statements about right and wrong that are separate from your own culture.

However, there are a number of problems with both these views. Against monism we can observe that it is simply a fact that people do hold very different moral values, not only between cultures but also between individuals. Against ethical relativism we can say that not only is it highly problematic to treat any culture as monolithic, but also that we need to be able to reach conclusions about ethical issues that enable us to act in the complex social world in which we live. At worst, relativism becomes 'a shrug of the shoulders and a "whatever"' (Hinman 2013: 34). Although it gives the appearance of creating tolerance and acceptance in conditions of diversity, ethical relativism leads either to the impossibility of action or to action that has some other basis. Because of the role and function of social work, inaction is not plausible. So, under ethical relativism we may expect to see the basis of practice and policy reduced to the stark exercise of power. Although social power is certainly exercised by social workers (Hugman 1991; Smith 2008), ethics serves as a mediating dimension to the relationships between social workers and service users, other professions, agencies and the state. For example, seeking the empowerment of service users is an ethical act (Adams 1990). To achieve this there needs to be a positive content to the ethics on which social work can draw.

Ethical pluralism is a different approach to monism and relativism. It recognises that all approaches to ethics have something to tell us about what is good and right and seeks to find a way to balance between different moral ideas. McBeath and Webb's (2002: 1019) complaint that social work ethics has become a 'persistent drone of Kantianism or utilitarianism or, out of the depths of the 1970s a mix of the two' can be seen as a form of pluralist argument, which seeks to connect virtue ethics with the two major strands of principle-based ethics that predominate in social work. In this sense contemporary social work ethics is not so much monism but dualism, as social work tends to assume that the two principle-based approaches fit easily together, or even are elements of the same set of values. The heart of McBeath and Webb's argument is that without a consideration of the ethical qualities and characteristics of practitioners, asserting principles in an absolute way lacks the capacity to address the way in which social work practice is accomplished in an uncertain and often messy world. It is a world of practice that requires social workers to exercise judgement and to make use of their moral compass, as well as following procedures or applying principles (McBeath and Webb 2002: 1033).

Like Banks and Gallagher (2009: 220), we argue that we need to take account of virtues alongside the principles of human dignity and rights and social justice in an ethical pluralist framework. For example, the social worker who has the qualities of humility and perseverance may be expected to approach their practice with people in different cultural communities with respect and dignity. Another example is the social worker who has the characteristics of courage and justice, who might be inclined to advocate strongly for disadvantaged or marginalised people. In both instances the goals of practice that are being pursued are grounded in the principles of human dignity and worth, human rights and social justice. The ethical virtues of the social worker give a shape to practice, arising from the moral character of social workers. In this sense we are suggesting that virtue ethics is central to the *practice* of social work.

From these reasons a pluralistic ethical framework is necessary in social work, to bring together the various approaches to ethics that are relevant (including not only virtues and the predominant principles, but also relationship-based and discursive approaches – for example, Hugman 2005a; Banks 2012). This is not an easy task, as a strength of principle-based approaches lies at least partly in their capacity to be used to create codes and other shared statements. Especially in neo-liberal systems the apparent objectivity of these types of statement is attractive, as it enables practices to be specified and risks calculated. Accepting the indeterminacy of ethical virtues in practice carries with it the sort of risks to which many contemporary structures and systems are highly averse. As McBeath and Webb observe, 'virtue theory under complexity shows a lack of firm boundary conditions' (2002: 1033). In other words, practice informed by ethical virtues cannot be predetermined or prescribed by organisational procedures in precise terms. This is also the reason why it would not be appropriate or helpful to try to specify a definitive list of ethical virtues that could be required of all social workers. Virtue ethics focuses on the person acting and not the objective status of the act (Hinman 2013: 46).

We are left, therefore, with the challenge of knowing what qualities or characteristics we should consider as the basis for someone becoming and being a good social worker. Following from that, we must also consider who is responsible for the development of these qualities and characteristics in social workers. The answer to these questions is effectively the same. The ethical virtues are those that are affirmed by social workers individually and collectively; the responsibility for developing and sustaining these virtues belongs with each social worker and with the profession as a whole.

To enable both individual social workers and the profession collectively to respond to the challenge of virtue ethics requires several things. First, virtue ethics needs to be addressed explicitly, in professional education and other formal training as well as in the ethical documents of professional bodies such as national associations. Many national codes of ethics and the international statement of ethical principles (IFSW/IASSW 2018) now contain sections on 'professional integrity' that, at least partly, address questions of the ethical qualities of social workers; they also tend to be focused on the actions of social workers as compared to the quality of the actors. Indeed, as we have acknowledged in this discussion, specifying ethical virtues of practitioners in codes is not feasible. For this reason, such documents are part of the greater recognition of virtue ethics in social work, but they are not the whole story.

Second, social workers could be assisted to think more about ethical virtues in their practice. Through our discussion we have noted that critical reflection plays an important role in ethical considerations. The use of critical reflection as a key element of social workers engaging in continuing professional development has become widely recognised in recent years (Gardner 2014; Pawar and Anscombe 2015). As part of practice, critical reflection enables social workers to consider their own values, the issues they face in their practice, and the organisational and political dimensions of practice (Gardner 2014: 129–134). This provides the opportunity to place an understanding of personal ethical qualities and characteristics in the wider context of practice. Likewise, supervision provides an opportunity to look more critically at oneself. However, in the pressures of everyday work there can be a tendency in supervision to focus on practical issues, decisions and so on. Gardner (2014: 107–108) notes that there are several approaches to supervision, including a managerial task-focused orientation, so that conscious effort needs to be made to consider ethical issues for the practitioner in terms of their own self as well as whether the work is being done in accordance with standards of conduct or some other external framework. Social work practitioners and supervisors together need to consider the ethics of practice as an important aspect of professional development.

Third, social workers need to overcome the tendency to see ethics as either a luxury, useful only when there is an 'ethical dilemma', or even of no use to the busy practitioner. Our own understanding of an ethical dilemma is that it represents a situation in which there is more than one way of acting that could be considered ethically appropriate, or that the choice facing the social worker is between solutions that are equally bad (Bowles *et al.* 2006: 42). Often the term ethical dilemma is used for situations that are complex or challenging but where the right course of action is apparent. However, ethical consideration helps in distinguishing better or worse ways to practise, as well as better or worse goals for practice. Thus, it can usefully draw on thinking about the ethical qualities of practitioners, including their moral compass seen through emotional responses to situations and the role this plays in practical wisdom (as discussed in Chapter 8).

Fourth, social workers need to overcome the innate sense that talking about ethics is to engage in moral judgementalism, especially when ethical questions are asked of specific individuals. While many social workers are able to be critical about policies and organisations (including politicians and senior managers, and perhaps even other agencies), there is less ease in thinking about specific instances at the practice level. Related to this is a diffidence that we have noted about the notion of 'ethical virtue' itself. In everyday conversation this is a challenging concept, as it is often understood as arrogance if applied to oneself (as if talking about oneself in relation to virtue is to be claiming moral perfection or otherwise setting oneself above others). To change this, social work may need to develop a new ethical language around such questions, in order to recognise that the way of talking comfortably about ethical virtues necessarily does not use the language of moral philosophy but is more descriptive about the parts of virtuous practice (thoughts, feelings, relationships, actions and so on). One of our hopes for this study is that it can contribute to the growth of dialogue around such issues.

Some concluding thoughts

More broadly, we aspire to this book adding to the debates about ethics in social work, so that virtues can be more clearly part of the pluralist ethical framework of the profession. Overcoming the neglect of the virtues has the potential to help social work to develop greater ethical depth. Whether involved in micro, meso or macro practices, all social workers bring their moral qualities and characteristics to the profession. So our goal for this book is to encourage a more conscious recognition of this dimension of our professional selves and the ways in which the qualities and characteristics that we bring to our roles have a significant impact on the way that others experience social work, as well as the ways in which we can seek to achieve the goals of the profession. The ethical virtues of social workers enable the values of human dignity and worth, human rights and social justice to be pursued in specific situations, in a variety of ways in the different fields and methods of practice.

Some of the virtues that we have identified are the same as those discussed in other studies, some are not addressed elsewhere, while also we do not consider some that others include in their lists (Bowles *et al.* 2006; Banks and Gallagher 2009; Banks 2012; Papouli 2018). This is inevitable, given the way in which each study has selected the virtues it addresses and the (perhaps subtle) shifts of emphasis over time. However, this variation contributes to the richness of our understanding because it adds to the dialogue that is necessary for ethical awareness and sensitivity to grow. Virtue ethics more than other approaches requires each person to be conscious of their own ethical qualities and characteristics. At the same time, communities of people who are attending consciously to ethical virtues in this way contribute to individuals' own development of such qualities and characteristics. This points to the collective responsibility of professional bodies and trade unions, agencies and other groups in social work to be actively engaged in the process. When this occurs, it will be possible for the profession to achieve a clearer view of what it is to be a good social worker.

FINAL REFLECTIVE QUESTIONS/EXERCISES

List three key learnings from this book.

Do you think virtue ethics can contribute usefully to your practice as a social worker?

The authors suggest that virtue ethics be further developed and cultivated in social workers through education, mentoring and modelling. Do you feel that virtue ethics has the potential to usefully inform and shape the social work profession? If so, how might you cultivate attention to virtues in your sphere of influence?

REFERENCES

Adams, P. (2009) Ethics with character: virtues and the ethical social worker, *Journal of Sociology & Social Welfare*, 36:3, 83–105.

Adams, R. (1990) *Self-Help, Social Work and Empowerment*. London: Macmillan.

Alexandra, A. & Miller, S. (2009) *Ethics in Practice: Moral Theory and the Professions*. Sydney: UNSW Press.

Alinsky, S. D. (1971) *Rules for Radicals*. New York: Random House.

Alston, M. (2013) Environmental social work: accounting for gender in climate disasters, *Australian Social Work*, 66:2, 218–233.

Androff, D. (2010) Truth and reconciliation commissions (TRCs): an international human rights intervention and its connection to social work, *British Journal of Social Work*, 40:6, 1960–1977.

Anthony, T. (2019) Fact check Q&A: are Indigenous Australians the most incarcerated people, on Earth?, *The Conversation*, electronic document downloaded on 8 August 2019 from https://theconversation.com/factcheck-qanda-are-indigenous-australians-the-most-incarcerated-people-on-earth-78528.

Armstrong, A. E. (2007) *Nursing Ethics: A Virtue-Based Approach*. Basingstoke: Palgrave Macmillan.

Arries, E. (2005) Virtue ethics: an approach to moral dilemmas in nursing, *Curationis*, 28:3, 64–72.

Australian Association of Social Workers [AASW] (2010) *AASW Code of Ethics*. Kingston, ACT: AASW.

Australian Council of Social Service [ACOSS] (2018) *Poverty in Australia 2018*. Strawberry Hills, NSW: ACOSS/UNSW.

Banks, S. (2004) *Ethics, Accountability and the Social Professions*. Basingstoke: Palgrave Macmillan.

Banks, S. (2010) Integrity in professional life: issues of conduct, commitment and capacity, *British Journal of Social Work*, 40:7, 2168–2184.

Banks, S. (2012) *Ethics and Values in Social Work* (4th edn). Basingstoke: Palgrave Macmillan.

Banks, S. (2016) Everyday ethics in professional life: social work as ethics work, *Ethics & Social Welfare*, 10:1, 35–52.

Banks, S. & Gallagher, A. (2009) *Ethics in Professional Life: Virtues for Health and Social Care.* Basingstoke: Palgrave Macmillan.
Barnes, M. (2012) *Care in Everyday Life: An Ethic of Care in Practice.* Bristol: Policy Press.
Bauman, Z. (1993) *Postmodern Ethics.* Oxford: Basil Blackwell.
Bauman, Z. (1994) *Alone Again: Ethics After Certainty.* London: Demos.
Bay, U. (2019) Neo-liberalism as an art of governance: reflecting on techniques for securing life through direct social work practice, *European Journal of Social Work*, 22:2, 201–211.
Beauchamp, T. L. & Childress, J. F. (2013) *Principles of Biomedical Ethics* (7th edn). New York: Oxford University Press.
Beckett, C. & Maynard, A. (2005) *Values and Ethics in Social Work: An Introduction.* London: Sage Publications.
Beddoe, L., Staniforth, B. & Fouché, C. (2019) 'Proud of what I do but often… I would be happier to say I drive trucks': ambiguity in social workers' self-perception, *Qualitative Social Work*, 18:3, 530–546.
Belcher, J. R. & Tice, C. (2013) Power and social work: a change in direction, *Journal of Progressive Human Services*, 24:1, 81–93.
Bennett, B. & Green, S. (eds) (2019) *Our Voices: Aboriginal Social Work* (2nd edn). London: Red Globe Press.
Bennett, B. (2019) The importance of Aboriginal history for practitioners. In Bennett, B. & Green, S. (eds), *Our Voices: Aboriginal Social Work* (2nd edn). London: Red Globe Press.
Berlin, I. (1992) *The Crooked Timber of Humanity: Chapters in the History of Ideas*, edited by Hardy, H. New York: Vintage Books.
Besthorn, F. (2002) Radical environmentalism and the ecological self, *Journal of Progressive Human Services*, 13:1, 53–72.
Bibus, A. (2013) Applying approaches from moral philosophy, especially virtue ethics, when facing ethical dilemmas in social work, *Journal of Czech & Slovak Social Work*, 139:6, 33–50 [in English].
Biestek, F. P. (1957) *The Casework Relationship.* London: George Allen & Unwin.
Bisman, C. (2004) Social work values: the moral core of the profession, *British Journal of Social Work*, 34:1, 104–123.
Bisman, C. (2014) *Social Work: Value-Guided Practice for a Global Society.* New York: Columbia University Press.
Bland, R., Renouf, N. & Tullgren, A. (2015) *Social Work Practice in Mental Health: An Introduction* (2nd edn). Crows Nest, NSW: Allen & Unwin.
Blum, L. A. (1994) *Moral Perception and Particularity.* Cambridge: Cambridge University Press.
Bowles, W., Collingridge, M., Curry, S. & Valentine, B. (2006) *Ethical Practice in Social Work: An Applied Approach.* St. Leonards, NSW: Allen & Unwin.
Bozalek, V. (2013) Reframing social work ethics through a political ethic of care and social justice lens. In Banks, S. (ed.), *Ethics.* Bristol: Polity Press.
Briar-Lawson, K., Martinson, K., Briar-Bonpane, J. & Zox, K. (2011) Child welfare, the media and capacity building, *Journal of Public Child Welfare*, 5:2–3, 185–199.
British Association of Social Workers [BASW] (2014) *The Code of Ethics for Social Work.* Birmingham: BASW.
Carr, D. (2007) Character in teaching, *British Journal of Educational Studies*, 55:4, 369–389.
Chadiha, L., Ingersoll-Drayton, B., Dunkle, R. E. & Balestrery, J. (2017) Forms and meanings of respect: aging mothers and adult daughters with mental illness, *Families in Society: The Journal of Contemporary Human Services*, 98:4, 319–327.
Chechak, D. (2015) Social work as a value-based profession: value conflict and implications for practitioners' self-concepts, *Journal of Social Work Ethics and Values*, 12:2, 41–48.

Clark, C. (2006) Moral character in social work, *British Journal of Social Work*, 36:1, 75–89.

Clark, C. (2011) Evidence-based practice and professional wisdom. In Bondi, L., Carr, D., Clark, C. & Clegg, C. (eds), *Towards Professional Wisdom*. Aldershot: Ashgate, 45–62.

Clifford, D. & Burke, B. (2009) *Anti-Oppressive Ethics and Values in Social Work*. Basingstoke: Palgrave Macmillan.

Cooke, S. & Carr, D. (2014) Virtue, practical wisdom and character in teaching, *British Journal of Educational Studies*, 62:2, 91–110.

Corcoran, J. & Pillay, V. (2009) A review of the research on solution-focused therapy, *British Journal of Social Work*, 39:2, 234–242.

Cordoba, P. S. (2017) Representations of social work in the Australian news media, *Australian Social Work*, 70:4, 453–464.

Cox, D., La Caze, M. & Levine, M. P. (2003) *Integrity and the Fragile Self*. Aldershot: Ashgate.

Crowder, R. & Sears, A. (2017) Building resilience in social workers: an exploratory study on the impacts of a mindfulness-based intervention, *Australian Social Work*, 70:1, 17–29.

Culham, T. & Bai, H. (2011) Emotional intelligence meets virtue ethics: implications for educators, *Journal of Thought*, 46:3–4, 25–43.

Davies, M. (1994) *The Essential Social Worker* (3rd edn). Aldershot: Ashgate.

Day, P. R. (1981) *Social Work and Social Control*. London: Tavistock.

Dean, R. G. (2001) The myth of cross-cultural competence, *Families in Society: The Journal of Contemporary Human Services*, 82:6, 623–630.

Dimitrijoska, S. & Ilievski, V. (2016) Ethics and value dilemmas in social work, *Revista de Asistenta Sociala*, 15:1, 49–58 [in English].

Dominelli, L. (2002) *Anti-Oppressive Social Work Theory and Practice*. Basingstoke: Palgrave Macmillan.

Dominelli, L. (2012) *Green Social Work*. Cambridge: Polity Press.

Donaldson, L. P. & Mayer, L. M. (2014) Justice as a core virtue for social work practice, *Social Work & Christianity*, 41:2/3, 207–231.

Doogan, C. M. (2016) *Review of Recommendation 9 of the Moss Review*. Canberra: Commonwealth of Australia.

Du Plooy, L., Harms, L., Muir, K., Martin, B. & Ingliss, S. (2014) 'Black Saturday' and its aftermath, *Australian Social Work*, 67:2, 274–284.

Duffy, F. & Healy, J. (2011) Social work with older people in a hospital setting, *Social Work in Health Care*, 50:2, 109–123.

England, H. (1983) *Social Work as Art*. London: George Allen & Unwin.

Ferguson, H. (2001) Social work, individualization and life politics, *British Journal of Social Work*, 31:1, 41–55.

Ferguson, I. & Lavalette, M. (2004) Beyond power and discourse: alienation and social work, *British Journal of Social Work*, 33:3, 297–312.

Finch, J. & Groves, D. (1980) Community care and the family: a case for equal opportunity?, *Journal of Social Policy*, 9:4, 487–511.

Flaskas, C. (2007) Holding hope and hopelessness: therapeutic engagements with the balance of hope, *Journal of Family Therapy*, 29:3, 186–202.

Flesakar, K. & Larsen, D. (2012) To offer hope you must have hope, *Qualitative Social Work*, 11:1, 61–79.

Foucault, M. (2000) *Ethics: Essential Works of Foucault 1954–1984, Volume 1*, edited by Rabinow, P., translated by Hurley, R. Harmondsworth: Penguin.

Gardner, F. (2014) *Being Critically Reflective: Engaging in Holistic Practice*. Basingstoke: Palgrave Macmillan.

Gilbert, T. & Powell, J. L. (2010) Power and social work on the United Kingdom: a Foucauldian excursion, *Journal of Social Work*, 10:1, 3–22.

Gilligan, C. (1982) *In a Different Voice*. Cambridge, MA: Harvard University Press.

Goldberg, G. & Elliott, J. (1980) Situational ethics for unethical situations, *Journal of Sociology & Social Welfare*, 7:4, 478–486.

Gough, J. & Spencer, E. (2014) Ethics in action: an exploratory survey of social workers' ethical decision making and value conflicts, *Journal of Social Work Values and Ethics*, 11:2, 23–40.

Graham, M. (2002) *Social Work and African-Centred World Views*. Birmingham: Venture Press.

Gray, M. (2005) Dilemmas of international social work: paradoxical processes of indigenisation, universalism and imperialism, *International Journal of Social Welfare*, 14:3, 231–238.

Gray, M. & Fook, J. (2004) The quest for a universal social work: some issues and implications, *Social Work Education*, 23:5, 625–644.

Gray, M., Plath, D. & Webb, S. A. (2009) *Evidence-Based Social Work: A Critical Stance*. Abingdon: Routledge.

Green, S. & Baldry, E. (2008) Building Indigenous Australian social work, *Australian Social Work*, 61:4, 389–402.

Groessl, J. (2017) Leadership in the field: fostering moral courage, *Journal of Social Work Values and Ethics*, 14:1, 72–79.

Halifax, J. (2012) A heuristic model of enactive compassion, *Support and Palliative Care*, 6:2, 228–235.

Hardina, D. (2004) Guidelines for ethical practice in community organization, *Social Work*, 49:4, 595–604.

Hardy, M. (2015) *Governing Risk: Care and Control in Contemporary Social Work*. Basingstoke: Palgrave Macmillan.

Havel, V. (2004) An orientation of the heart. In Loeb, P. R. (ed.), *The Impossible Will Take a Little While*. New York: Basic Books.

Hay, J. (2019) 'Care is not a dirty word!': enacting an ethic of care in social work practice, *European Journal of Social Work*, 22:3, 365–375.

Held, V. (1993) *Feminist Morality: Transforming Culture, Society and Politics*. Chicago, IL: University of Chicago Press.

Hinman, L. M. (2013) *Ethics: A Pluralistic Approach*. Boston, MA: Wadsworth.

Hodge, D. R. (2002) Does social work oppress Evangelical Christians? A 'new class' analysis of society and social work, *Social Work*, 47:4, 401–414.

Hoff, M. D. & Polack, R. J. (1993) Social dimensions of the environmental crisis: challenges for social work, *Social Work*, 38:2, 204–211.

Hölscher, D. & Bozalek, V. (2012) Encountering the *Other* across the divides: re-grounding social justice as a guiding principle for social work with refugees and other vulnerable groups, *British Journal of Social Work*, 42:6, 1093–1112.

Houston, S. (2003) Establishing virtue in social work: a response to McBeath and Webb, *British Journal of Social Work*, 33:6, 819–824.

Houston, S. (2011) Engaging with the crooked timber of humanity: value pluralism and social work, *British Journal of Social Work*, 42:4, 652–668.

Howe, D. (2011) *Attachment Across the Life-Course: A Brief Introduction*. Basingstoke: Palgrave Macmillan.

Howe, D. (2014) *The Compleat Social Worker*. Basingstoke: Palgrave Macmillan.

Hugman, R. (1991) *Power in Caring Professions*. London: Macmillan.

Hugman, R. (2005a) *New Approaches in Ethics for the Caring Professions*. Basingstoke: Palgrave Macmillan.

Hugman, R. (2005b) Exploring the paradox of teaching ethics in social work, *Social Work Education*, 24:5, 535–545.

Hugman, R. (2013) *Culture, Values and Ethics in Social Work*. London: Routledge.

Hugman, R. (2017) Power and authority in social work practice: some ethical issues. In Hugman, R. & Carter, J. (eds), *Rethinking Values & Ethics in Social Work*. London: Palgrave.

Hume, D. (1966/1751) *An Enquiry Concerning the Principles of Morals*. La Salle, IL: Open Court Publishing.

Hunter, S. & Kendrick, M. (2009) The ambiguities of professional and societal wisdom, *Ethics & Social Welfare*, 3:2, 158–169.

Hursthouse, R. (1999) *On Virtue Ethics*. Oxford: Oxford University Press.

Husband, C. (1995) The morally active practitioner and the ethics of antiracist social work. In Hugman, R. & Smith, D. (eds), *Ethical Issues in Social Work*. London: Routledge.

Hyslop, I. (2018) Liberalism and social work identity, *European Journal of Social Work*, 21:1, 20–31.

Ife, J. (2012) *Human Rights and Social Work: Towards Rights-Based Practice* (3rd edn). South Melbourne, VIC: Cambridge University Press.

International Association of Schools of Social Work & International Federation of Social Workers [IASSW/IFSW] (2004) *Global Standards for Social Work Education and Training*, electronic document downloaded on 10 July 2019 from www.iassw-aiets.org/wp-content/uploads/2018/08/Global-standards-for-the-education-and-training-of-the-social-work-profession.pdf.

International Federation of Social Workers [IFSW] (2002) *Minutes of the IFSW General Meeting 2002*, electronic document downloaded on 7 August 2019 from www.ifsw.org/minutes-of-the-ifsw-general-meeting-2002/.

International Federation of Social Workers [IFSW] (2009) *Murder of Former PASWI President and Philippine Welfare Attaché*, electronic document downloaded on 7 August 2019 from www.ifsw.org/murder-of-former-paswi-president-and-philippine-welfare-attache/.

International Federation of Social Workers & International Association of Schools of Social Work [IFSW/IASSW] (2018) *Global Social Work Statement of Ethical Principles*, electronic document downloaded on 10 July 2019 from www.ifsw.org/global-social-work-statement-of-ethical-principles/.

Keith-Lucas, A. (1972) *The Giving and Taking of Help*. Chapel Hill, NC: University of North Carolina Press.

Kenny, S. & Connors, P. (2017) *Developing Communities for the Future* (5th edn). South Melbourne, VIC: Cengage Learning.

Kiley, K. A., Sehgal, A. R., Neth, S., Dolata, J., Pike, E., Splisbury, J. C. & Albert, J. M. (2018) Effectiveness of guided imagery in treating compassion fatigue and anxiety of mental health workers, *Social Work Research*, 42:1, 33–43.

Kittay, E. F. (1999) *Love's Labour: Essays on Women, Equality and Dependency*. New York: Routledge.

Koehn, D. (1994) *The Ground of Professional Ethics*. London: Routledge.

Koenig, T. & Spano, R. (2007) The cultivation of social workers' hope in personal life and professional practice, *Journal of Religion and Spirituality in Social Work: Social Thought*, 26:3, 45–61.

Kvarnstrom, S., Willumsen, E., Andersson-Gare, B. & Hedberg, B. (2012) How service users perceive the concept of participation, specifically in interprofessional practice, *British Journal of Social Work*, 42:1, 129–146.

Lai, K. (2006) *Learning from Chinese Philosophies: Ethics of Interdependent and Contextualised Self.* Aldershot: Ashgate.

Lavalette, M. & Ioakimidis, V. (eds) (2011) *Social Work in Extremis: Lessons for Social Work Internationally*. Bristol: Policy Press.

Levenson, J. (2017) Trauma informed social work practice, *Social Work*, 62:2, 105–113.

Lewis, C. S. (1998/1942) *The Screwtape Letters*. London: Fount/Harper Collins.

Loeb, P. R. (2004) Hope in a time of fear, *Nation*, 279:8, 22–28.

Lundy, C. (2006) *Social Work and Social Justice*. Calgary, AB: Broadview Press.

Luntz, J. (1991) Vision to reality: care study of a unique community mental health project, *Australian Social Work*, 44:2, 41–48.

Lymbery, M. (2001) Social work at the crossroads, *British Journal of Social Work*, 31:3, 369–384.

Lynch, D. & Forde, C. (2010) 'Moral distress' and the beginning practitioner: preparing social work students for ethical and moral challenges in contemporary contexts, *Ethics & Social Welfare*, 10:2, 94–107.

Macdonald, G., Sheldon, B. & Gillespie, J. (1992) Contemporary studies of the effectiveness of social work, *British Journal of Social Work*, 22:6, 615–643.

MacIntyre, A. (1985) *After Virtue: A Study in Moral Theory* (2nd edn). London: Duckworth.

MacIntyre, A. (1989) *A Short History of Ethics*. London: Routledge.

Marston, G. & McDonald, C. (2012) Getting beyond 'heroic agency' in conceptualising social workers as policy actors in the twenty-first century, *British Journal of Social Work*, 42:6, 1022–1038.

Mayer, J. E. & Timms, N. (1970) *The Client Speaks: Working Class Impressions of Casework*. London: Routledge & Kegan Paul.

Mbiti, J. (1990) *African Religions and Philosophy* (2nd edn). Oxford: Heinemann.

McAuliffe, D. (2005) I'm still standing: impacts and consequences of ethical dilemmas for social workers in direct practice, *Journal of Social Work Values and Ethics*, 2:1, 43–62.

McAuliffe, D., Williams, C. & Briskman, L. (2010) Moral outrage! Social work and social welfare, *Ethics & Social Welfare*, 10:2, 87–93.

McBeath, G. & Webb, S. A. (2002) Virtue ethics and social work: being lucky, realistic, and not doing your duty, *British Journal of Social Work*, 32:8, 1015–1036.

Midgley, J. (1981) *Professional Imperialism*. London: Heinemann.

Moller Okin, S. (1989) *Justice, Gender and the Family*. New York: Basic Books.

Morales-Sánchez, R. & Cabello-Medina, C. (2013) The role of four universal moral competencies in decision-making, *Journal of Business Ethics*, 116:4, 717–734.

Morse, J. (1999) The missing link between virtue theory and business ethics, *Journal of Applied Philosophy*, 16:1, 47–58.

Moss, P. (2015) *Review into Recent Allegations Relating to Conditions and Circumstances at the Regional Processing Centre in Nauru*. Canberra: Commonwealth of Australia.

Musschenga, A. W. (2001) Education for moral integrity, *Journal of Philosophy of Education*, 35:2, 219–235.

Nagel, T. (1979) *Mortal Questions*. Cambridge: Cambridge University Press.

Newham, R. A. (2015) Virtue ethics and nursing: on what grounds?, *Nursing Philosophy*, 16:1, 40–50.

Noddings, N. (1984) *Caring: A Feminine Approach to Ethics and Moral Education*. Berkeley, CA: University of California Press.

Nussbaum, M. C. (2001) *Upheavals of Thought: The Intelligence of Emotions*. New York: Cambridge University Press.

Oakley, J. (1993) *Morality and the Emotions*. London: Routledge.

Oakley, J. & Cocking, D. (2001) *Virtue Ethics and Professional Roles*. Cambridge: Cambridge University Press.

O'Brien, M. (2011) Social justice: alive and well (partly) in social work practice? *International Social Work*, 54:2, 174–190.

Olver, I. N. (2012) Evolving definitions of hope in oncology, *Current Opinion in Supportive and Palliative Care*, 6:2, 236–241.

Papouli, E. (2016) Teaching and learning for ethical practice in social work education. In Taylor, I., Bogo, M., Lefevre, M. & Teater, B. (eds), *Routledge International Handbook of Social Work Education*. Abingdon: Routledge.

Papouli, E. (2018) Aristotle's virtue ethics as a conceptual framework for the study and practice of social work in modern times, *European Journal of Social Work*, DOI: 10.1080/13691457.2018.1461072.

Papouli, E. (2019a) Diversity dolls: a creative teaching method for encouraging social work students to develop empathy and understanding for vulnerable populations, *Social Work Education*, 38:2, 241–260.

Papouli, E. (2019b) Moral courage and moral distress in social work education and practice. In Marson, S. & McKinney, R. (eds), *The Routledge Handbook of Social Work Ethics and Values*. New York: Routledge.

Parrott, L. (2010) *Values and Ethics in Social Work Practice*. Exeter: Learning Matters Limited.

Parrott, L. (2014) *Values and Ethics in Social Work Practice* (3rd edn). Exeter: Learning Matters.

Pawar, M. & Anscombe, B. (2015) *Reflective Social Work Practice: Thinking, Doing and Being*. Melbourne, VIC: Cambridge University Press.

Pawar, M., Hugman, R., Alexandra, A. & Anscombe, A. W. (eds) (2017) *Empowering Social Workers: Virtuous Practitioners*. Singapore: Springer.

Pawar, M., Hugman, R., Anscombe, A. W. & Alexandra, A. (2019) Searching for virtue ethics: a survey of social work ethics curriculum and educators, *British Journal of Social Work*, DOI: 10.1093/bjsw/bcz106.

Payne, M. (2005) *The Origins of Social Work: Continuity and Change*. Basingstoke: Palgrave Macmillan.

Pellegrino, E. (2005) Some things ought never to be done: moral absolutes in clinical ethics, *Theoretical Medicine and Bioethics*, 26:6, 469–486.

Pellegrino, E. (2007) Professional medicine, virtue-based ethics and the retrieval of professionalism. In Walker, R. L. & Ivanhoe, P. J. (eds), *Working Virtue: Virtue Ethics and Contemporary Moral Problems*. New York: Oxford University Press.

Pellegrino, E. D. & Thomasma, D. C. (1993) *The Virtues in Medical Practice*. Oxford: Oxford University Press.

Peterson, C. & Seligman, M. E. P. (2004) *Character Strengths and Virtues: A Handbook and Classification*. New York: Oxford University Press.

Phillips, N. & Hugman, R. (1999) The users' perspective: the experience of being a parent with a mental health problem. In Weir, A. & Douglas, A. (eds), *Child Protection and Adult Mental Health: Conflict of Interest?* London: Butterworth-Heinemann.

Pines, A. & Aronson, E. (1983) Combatting burnout, *Children and Youth Services Review*, 5:3, 263–275.

Plath, D. (2014) Implementing evidence-based practice: an organisational perspective, *British Journal of Social Work*, 14:4, 905–923.

Postle, K. & Beresford, P. (2007) Capacity building and the reconceptualisation of political participation: a role for social care workers? *British Journal of Social Work*, 37:1, 143–158.

Pullen-Sansfaçon, A. (2010) Virtue ethics for social work: a new pedagogy for practical reasoning, *Social Work Education*, 29:4, 402–415.

Pullen-Sansfaçon, A. & Cowden, S. (2013) *The Ethical Foundations of Social Work*. Abingdon: Routledge.

Pyles, L. (2019) Relational ethics and transformative community organising in the US. In Banks, S. & Westoby, P. (eds), *Ethics, Equity and Community Development*. Bristol: Policy Press.

Rawls, J. (1971) *A Theory of Justice*. Cambridge, MA: Harvard University Press.

Reamer, F. G. (2001) *Ethics Education in Social Work*. Alexandria, VA: Council on Social Work Education.

Reamer, F. G. (2003) Social work, Evangelical Christians and values, *Social Work*, 48:3, 428–431.

Reamer, F. G. (2018) *Social Work Values and Ethics* (5th edn). New York: Columbia University Press.

Rees, S. (1991) *Achieving Power: Practice and Policy in Social Welfare*. North Sydney, NSW: Allen & Unwin.

Rees, S. & Wallace, A. (1982) *Verdicts on Social Work*. London: Edward Arnold.

Reichert, E. (2017) Human rights and ethics. In Hugman, R. & Carter, J. (eds), *Rethinking Values & Ethics in Social Work*. London: Palgrave.

Reisch, M. (2017) Social justice. In Hugman, R. & Carter, J. (eds), *Rethinking Values & Ethics in Social Work*. London: Palgrave.

Rhodes, M. L. (1986) *Ethical Dilemmas in Social Work Practice*. Boston, MA: Routledge & Kegan Paul.

Robinson, D. (1999) *Nietzsche and Postmodernism*. Cambridge/New York: Icon Books/ Totem Books.

Russ-Smith J. (2019) Indigenous social work and a Wiradyuri framework to practice. In Bennett, B. & Green, S. (eds), *Our Voices: Aboriginal Social Work* (2nd edn). London: Red Globe Press.

Sacco, T. & Hoffmann, W. (2004) Seeking truth and reconciliation in South Africa, *International Social Work*, 47:2, 157–167.

Saini, M., Black, T., Lwin, K., Marshall, A., Fallon, B. & Goodman, D. (2012) Child protection workers' experiences of working with high-conflict separating families, *Children and Youth Services Review*, 34:7, 1309–1316.

Sainsbury, E. E. (1975) *Social Work with Families*. London: Routledge & Kegan Paul.

Sainsbury, E., Nixon, S. & Phillips, D. (1982) *Social Work in Focus: Clients' and Social Workers' Perceptions of Long-Term Social Work*. London: Routledge & Kegan Paul.

Saleebey, D. (ed.) (2002) *The Strengths Perspective in Social Work Practice*. Boston, MA: Allyn & Bacon.

Saleebey, D. (2004) The 'power of place': another look at the environment, *Families in Society: the Journal of Contemporary Social Services*, 85:1, 7–16.

Sandel, J. (2011) Justice: what's the right thing to do?, *Boston University Law Review*, 91:4, 1303–1310.

Sellman, D. (1996) Why teach ethics to nurses?, *Nurse Education Today*, 16:1, 44–48.

Sercombe, H. (2010) *Youth Work Ethics*. London: Sage Publications.

Sevenhuijsen, S. (1998) *Citizenship and the Ethics of Care: Feminist Considerations on Justice, Morality and Politics*. London: Routledge.

Sewpaul, V. (2016) The West and the Rest divide: human rights, culture and social work, *Journal of Human Rights Social Work*, 1:1, 30–39.

Shun, K.-L. & Wong, D. B. (eds) (2004) *Confucian Ethics: A Comparative Study of Self, Autonomy and Community*. Cambridge: Cambridge University Press.

Slote, M. (1992) *From Morality to Virtue*. New York: Oxford University Press.

Slote, M. (2001) *Morals from Motives*. Oxford: Oxford University Press.

Slote, M. (2007) *The Ethics of Care and Empathy*. London: Routledge.

Slote, M. (2011) *The Impossibility of Perfectionism: Aristotle, Feminism and the Complexities of Ethics*. New York: Oxford University Press.

Smith, D. (ed.) (2004) *Social Work and Evidence Based Practice*. London and Philadelphia, PA: Jessica Kingsley.

Smith, R. (2008) *Social Work and Power*. Basingstoke: Palgrave Macmillan.

Solas, J. (2008) Social work and social justice: what are we fighting for?, *Australian Social Work*, 61:2, 124–136.

Solomon, R. (1992) *Ethics and Excellence*. New York: Oxford University Press.

Steen, J. A., Mann, M., Restivo, N., Mazeny, S. & Chapple, R. (2016) Human rights: its meaning and practice in field settings, *Social Work*, 62:1, 9–17.

Strike, K. A. & Soltis, J. F. (2015) *The Ethics of Teaching* (5th edn). New York: Teachers' College Press.

Sullivan, B., Grant, S. & Grant, F. (2016) *Yindyamarra Yambuwan, Creative Works – Recorded or Rendered Works*. Wagga Wagga/Dubbo, NSW: Charles Sturt University.

Thyer, B. (2013) Evidence-based practice or evidence-guided practice: a rose by any other name would smell as sweet, *Families in Society*, 94:2, 79–84.

Timms, N. (1983) *Social Work Values: An Inquiry*. London: Routledge & Kegan Paul.

Titmuss, R. (1971) *The Gift Relationship*. London: George Allen & Unwin.

Trappenburg, M. & van Beek, G. (2019) 'My profession is gone': how social workers experience de-professionalization in the Netherlands, *European Journal of Social Work*, 22:4, 676–689.

Trevithick, P. (2012) *Social Work Skills and Knowledge: A Practice Handbook*. Maidenhead: Open University Press.

Tronto, J. (1993) *Moral Boundaries: A Political Argument for an Ethic of Care*. New York: Routledge.

Truax, C. & Carkhuff, R. (1967) *Towards Effective Counseling and Psychotherapy: Training and Practice*. Chicago, IL: Aldine Publishing Company.

Tutu, D. (1999) *No Future Without Forgiveness*. New York: Doubleday.

Ungar, M. (2008) Resilience across cultures, *British Journal of Social Work*, 38:2, 218–235.

Ungerson, C. (1987) *Policy is Personal*. London: Tavistock.

United Nations (1951) *Convention Relating to the Status of Refugees*. New York: United Nations.

van Hooft, S. (1995) *Caring: An Essay in the Philosophy of Ethics*. Niwot, CO: University Press of Colorado.

van Hooft, S. (2014) *Understanding Virtue Ethics*. Abingdon: Routledge.

van Wormer, K. (2003) Restorative justice: a model for social work practice with families, *Families in Society*, 84:3, 441–448.

Voss, R. W., Douville, V., Alex Little Soldier & Twiss, G. (1999) Tribal and Shamanic-based social work practice: a Lakota perspective, *Social Work*, 44:3, 228–241.

Walker, S. (2012) The teaching of Mäori social work practice and theory to a predominantly Päkehä audience, *Aotearoa New Zealand Social Work*, 24:3–4, 65–74.

Walsh, F. (2016) Applying a family resilience framework in training, practice and research, *Family Process*, 55:4, 616–632.

Warner, J. (2014) 'Heads must roll?' Emotional politics, the press and the death of Baby P., *British Journal of Social Work*, 44:6, 1637–1653.

Weaver, H. (1999) Indigenous peoples and the social work profession: defining culturally competent services, *Social Work*, 44:3, 217–225.

Weaver, H. (2017) Ethics and settler societies. In Hugman, R. & Carter, J. (eds), *Rethinking Values & Ethics in Social Work*. London: Palgrave.

Webb, S. A. (2006) *Social Work in a Risk Society: Social and Political Perspectives*. Basingstoke: Palgrave Macmillan.

Webb, S. A. (2010) Virtue ethics. In Gray, M. & Webb, S. A. (eds), *Ethics and value Perspectives in Social Work*. Basingstoke: Palgrave Macmillan.

Weinberg, M. (2009) Moral distress: a missing but relevant concept for ethics in social work, *Canadian Social Work Review*, 26:2, 139–151.

Weinberg, M. (2010) The social construction of social work ethics: politicizing and broadening the lens, *Journal of Progressive Human Services*, 21:1, 32–44.

Williams, B. (1981) *Moral Luck: Philosophical Papers 1973–1980*. Cambridge: Cambridge University Press.

Worsley, L. (2015) The resilience doughnut: combining strengths to survive. In Patron, M.-C. & Holden, S. S. (eds), *Victim Victorious: from Fire to Phoenix*. New York: Nova Science Publishers.

Wronka, J. (1998) *Human Rights and Social Policy in the 21st Century* (revised edn). Lanham, MD: University Press of America.

Yip, K.-S. (2004) A Chinese cultural critique of the global qualifying standards for social work education, *Social Work Education*, 23:5, 597–612.

Young, M. I. (1990) *Justice and the Politics of Difference*. Princeton, NJ: Princeton University Press.

Zimbardo, P. (2007) *The Lucifer Effect: How Good People Turn Evil*. New York: Random House.

INDEX